A sustainability checklist for developments

A common framework for developers and local authorities

Deborah Brownhill
Susheel Rao

BRE Centre for Sustainable Construction

BRE
Garston
Watford
WD25 9XX

Prices for all available
BRE publications can be
obtained from:
CRC Ltd
151 Rosebery Avenue
London EC1R 4GB
Tel: 020 7505 6622
Fax: 020 7505 6606
email:
crc@construct.emap.co.uk

BR 436
ISBN 1 86081 533 2

© Copyright BRE 2002
First published 2002

BRE is committed to providing impartial and authoritative information on all aspects of the built environment for clients, designers, contractors, engineers, manufacturers, occupants, etc. We make every effort to ensure the accuracy and quality of information and guidance when it is first published. However, we can take no responsibility for the subsequent use of this information, nor for any errors or omissions it may contain.

Published by
Construction Research
Communications Ltd
by permission of
Building Research
Establishment Ltd

Requests to copy any part of this publication should be made to:
CRC Ltd, Building Research Establishment, Bucknalls Lane, Watford WD25 9XX

Printed on paper produced from timber grown in sustainably managed forests

The views expressed in this document are not necessarily those of the Secretaries of State for Trade and Industry and for Transport, Local Government and the Regions.

BRE material is also published quarterly on CD

Each CD contains BRE material published in the current year, including reports, specialist reports, and the Professional Development publications: Digests, Good Building Guides, Good Repair Guides and Information Papers.

The CD collection gives you the opportunity to build a comprehensive library of BRE material at a fraction of the cost of printed copies.

As a subscriber you also benefit from a 20% discount on other BRE titles.

For more information contact:
CRC Customer Services on 020 7505 6622

Construction Research Communications

CRC supplies a wide range of building and construction related information products from BRE and other highly respected organisations.

Contact:
post: CRC Ltd
        151 Rosebery Avenue
        London EC1R 4GB

fax: 020 7505 6606
phone: 020 7505 6622
email: crc@construct.emap.co.uk
website: www.constructionplus.co.uk

Contents

Preface	v
Acknowledgements	vi

Introduction   1
   Purpose of the checklist   1
   Form of the checklist   1
   The issue headings   2
   Questions   2
   Range of performance   2
   Suggested maximum scores   3
   Ways in which this checklist can be used   3
   The principles underlying the issues: understanding the context   4

Completing the checklist questions   7

1 Land use, urban form and design   9
   1.1 Site criteria   10
   1.2 Reusing sites   12
   1.3 Form of development: grain, layout, scale   16
   1.4 Open space/Landscaping   18
   1.5 Density   18
   1.6 Mix of uses   20
   1.7 Aesthetic aspects   22

2 Transport   25
   2.1 General policy   26
   2.2 Public transport provision   28
   2.3 Parking   30
   2.4 Facilities for pedestrians and cyclists   32
   2.5 Provision of local employment   32
   2.6 Proximity of local facilities   34

3 Energy   37
   3.1 Community-wide energy production   38
   3.2 Street lighting   40

4 Impact of individual buildings   43
   4.1 Meeting a specified BREEAM rating   44
   4.2 Building types not covered by BREEAM   48
      (a) CO_2 emissions targets   48
      (b) Use of green materials   52
      (c) Water targets   54
      (d) Health and wellbeing   54
      (e) Transport   56
      (f) Pollution   56

Contents

5 Natural resources — 59
    5.1 Use of locally reclaimed/green materials — 60
    5.2 Air quality — 62
    5.3 Water conservation — 62
    5.4 Sustainable drainage — 64
    5.5 Refuse composting — 66

6 Ecology — 69
    6.1 Conservation — 70
    6.2 Enhancement of existing site — 70
    6.3 Planting — 72

7 Community — 75
    7.1 Community involvement and identity — 76
    7.2 Measures taken to reduce the opportunity for crime — 78

8 Business — 81
    8.1 Enhanced business opportunities — 82
    8.2 Employment and training — 84

Calculating the overall scores for social, environmental and economic sustainability — 86

Weighting and ranking of the issues — 88

Preface

How can we ensure that our towns and cities are developed and regenerated to be sustainable for the future? This is one of the greatest challenges facing those involved in planning and development.

In the *UK Strategy for Sustainable Development*[1], the Government outlined four underlying themes that describe the sustainability objectives for development. These themes, which appear in all levels of guidance for the construction industry (from planning through design to development), are:
- maintenance of high and stable levels of economic growth and employment,
- social progress which recognises the needs of everyone,
- effective protection of the environment, and
- prudent use of natural resources.

The Urban Task Force and the UK Round Table on Sustainable Development suggest the use of indicators to monitor progress towards sustainability. The Urban White Paper[2] takes a holistic view when tackling urban decline by urging that, in order to improve prosperity and quality of life, issues such as education, transport and crime need to be addressed alongside construction and planning issues. The *National strategy on neighbourhood renewal*[3] reiterated this stance.

Considerable ongoing research (at BRE[4] and elsewhere) has found it difficult to define appropriate indicators due to the broad and complex nature of sustainability. However, this research shares a common view that in order to be meaningful, indicators must act at the appropriate level (eg national indicators are often not appropriate at the local level, etc.). Consistent with this, BRE studies[5] involving local authorities highlight the need for practical tools and indicators to measure the sustainability of developments (both buildings and infrastructure) at site or estate level. This checklist provides these tools and can be used by developers, designers, planners and the community as a common framework for discussion. We hope to refine and develop this approach both as new knowledge becomes available and in response to feedback from users. Please address your comments and suggestions to: Deborah Brownhill or Susheel Rao, Centre for Sustainable Construction, BRE, Bucknalls Lane, Watford WD25 9XX.

DB, SR
2002

References

[1] **DETR.** *UK strategy for sustainable development.* London, The Stationery Office. 2000.
[2] **DETR.** *Our towns and cities: the future. Delivering an urban renaissance.* Government White Paper. London, The Stationery Office. 2000.
[3] **Cabinet Office Social Exclusion Unit.** *A new commitment to neighbourhood renewal: national strategy action plan.* London, The Stationery Office. 2001.
[4] **DETR Foresight Panel: Sustainability Indicators for Construction.** Sustainability indicators for the Movement for Innovation.
[5] **Crowhurst D, Rao S, McAllister I.** *Local authorities' performance on sustainable construction.* BRE Information Paper IP7/01. Garston, CRC. 2001.

Acknowledgements

This guide is the result of a project sponsored by the Department of Trade and Industry with Industry collaboration under the Partners in Technology programme, and supported by a committee of planners and developers made up of the following members:

Jon Muncaster	English Partnerships
Colin Percy	Newcastle City Council
Andrew Leno	Watford Council
Ian Lindley	Leicester City Council
Andrew Dutton	Persimmon, Bryant and Taywood Homes
Robert Tomlinson	Living Village Trust
Cressida Phillips	Nightingale Associates
Derry Caleb	University of Surrey
Martin Crookston	Lord Roger's Urban Task Force/Llewelyn Davies Planning Consultants
Anna Burgess	Hertfordshire County Council

Other organisations consulted were Buro Happold and WSP Consultants. BRE staff consulted were Alan Yates, Nigel Howard, Ian Dickie, Suzy Edwards, Richard Hartless, Matthew Ling, Jane Anderson and Helen Sargant.

Introduction

Purpose of the checklist

This checklist is designed to be used by those involved in planning or building sizeable developments from estates to urban villages and regeneration projects. It helps at the detailed estate/site level, focussing on the sustainability aspects relating to buildings and infrastructure. Using it will:
- increase awareness amongst urban planners, developers and estate managers of the practical measures that can be taken to plan 'sustainability' into a development,
- provide a framework for assessing the sustainability issues relating to buildings and infrastructure,
- give guidance on standards and indicators,
- be capable of adaptation to the local circumstances,
- provide developers with a method of demonstrating to planning authorities that sustainability has been systematically addressed in their proposals,
- help planners to specify 'sustainability' in supplementary planning guidance/development codes, etc.,
- provide planners with a method of assessing the sustainability aspects of development proposals consistent with DTLR requirements.

When considering the 'sustainability' of a whole area or community (rather than the impacts of individual buildings), social and economic issues, crucial to the vitality of a community, need to be included alongside environmental issues. Social and economic problems vary greatly from area to area as do the answers to these problems. This checklist is presented as a significant first step to help assess all three of these factors in proposed developments.

Form of the checklist

The checklist has been designed to be flexible in the way(s) it can be used, reflecting the practical feedback received from users to date. It is tailored towards application in the urban environment, mostly at the local planning level, and may be incorporated (either in whole or part) in Supplementary Planning Guidance/Development Codes/Development Briefs. It is intended to be used in support of the sustainability aims in the Local/Unitary Development Plan and to reflect the requirements of Planning Policy Guidance Notes and DTLR and DTI Good Practice Guides.

The checklist works by considering the positive measures that can be taken to reduce environmental impact or enhance social and economic benefits. It encourages good practice by merits in the scoring system (it does not penalise poor practice by deducting marks). The four main parts of the checklist are as follows.

Part 1 Questions	Part 2 Answering the Questions	Part 3 Range of performance	Part 4 Suggested maximum scores for achieving Best Practice
These are questions to assess specific issues that affect sustainability.	Explanation is given of how to answer the questions and score them.	Each issue is given Minimum, Good and Best Practice standards.	Each issue has a suggested maximum scoring value presented in three parts reflecting Environmental (E), Social (S) and Economic (Ec) impacts.

The issue headings

For convenience, a set of eight issue headings that encompass the underlying sustainability principles have been defined. The principles are described later. The eight issue headings are as follows.

- Land use, urban form and design
- Transport
- Energy
- Buildings
- Natural resources
- Ecology
- Community
- Business

The issue headings are not mutually exclusive as sustainability issues are often interrelated. Beneath each of the issue headings are a number of sub issues, for example, 'Intensity of land use' and 'Mix of uses' are two sub issues within *Land use, urban form and design*.

Questions

For each sub issue, specific questions have been designed to cover the main aspects of sustainability. For example, the sub issue 'Intensity of land use' has one question relating to the density of the housing which is measured in dwellings per hectare.

Range of performance

For each of the sub issues, the checklist provides suggested ranges of performance or standards. These range from 'Minimum acceptable', through 'Good Practice' to 'Best Practice'. The ranges of performance in this checklist are derived from current planning guidance notes/good practice guidance or based on scientific research, wherever possible. Where good practice guidance is not yet available, standards were agreed by a consensus view of the committee, and on the experience of testing this checklist on six different schemes during this project.

When setting standards, experience from the BRE Environmental Assessment Method (BREEAM) has illustrated the need to be flexible; often, demanding 100% compliance with a standard is an impossible target and can lead to an issue being ignored completely, rather than partially addressed. In some cases, therefore, this checklist adopts an approach called the 80/20 rule. This means that Best Practice can be achieved provided more than 80% of the development meets the particular standard.

The checklist standards have been derived for 'typical' developments. However, as no development is ever 'typical' in all regards, these values should be checked for appropriateness whenever the checklist is used. For example, the standards for housing density in dwellings per hectare (DPH) are:

- Minimum acceptable 35–40 DPH,
- Good Practice 41–59 DPH,
- Best Practice ≥ 60 DPH.

However, housing density is correctly affected by the size and scale of the surrounding development and also by proximity to public transport. These other factors must be taken into consideration and the standard revised accordingly.

Suggested maximum scores for achieving Best Practice

Against each question and performance range, the checklist has a recommended set of maximum scores reflecting the Environmental, Social and Economic benefits of achieving Best Practice. For questions where a given aspect is not relevant, there is no score.

For example, another question under the sub issue 'Intensity of land use' relates to the number of homes with easy access to public green space. Access to green space is considered to bring both social and environmental benefits, but not necessarily economic benefits (unless local factors change this). The recommended maximum score in the economic category for this sub issue is therefore zero.

BRE has developed a strong track record in quantifying environmental impacts. However, assigning values for sustainability (especially social and economic aspects) cannot be hard science. One reason for this is that the social and economic value of development to a region is very much dependent on the starting position and local priorities. The values given are for a 'typical' case but as no case is ever 'typical' the scores can be altered and weighted differently by the user to reflect local circumstances.

The relative importance given to the different issues by BRE, across the spectrum of social, environmental and economic sustainability, is based both on BRE's research into the impacts of construction, and the results of detailed consensus research designed to understand how different organisations and individuals rank key sustainability issues. For a more detailed description of the weighting and ranking of the key sustainability issues see *Weighting and ranking of the issues* on page 88, and for a description of this research see BRE *Digest 446*.

Ways in which this checklist can be used

The checklist is flexible. Parts or all of it can be used as appropriate. Judgement is required on the part of the user when tailoring the checklist to reflect the local circumstances. Some of the ways that the checklist might be used are listed below.

- **Writing development briefs or proposals**
  The first two parts of the checklist (Questions and Answering the Questions) can be used together as a tick box system or *aide-mémoire* to ensure that sustainability has been thoroughly considered.
- **Demonstrating the sustainable aspects of a development proposal**
  The first three parts (Questions, Answering the Questions and Range of performance) can be used together by a developer to demonstrate the sustainability qualities of a proposal.
- **To specify sustainability standards**
  The first three parts (Questions, Answering the Questions and Range of performance) can be used together by a planning authority, land owner or other similar body to specify particular standards or targets to meet.
- **To perform a numerical evaluation of development proposals**
  All four parts of the checklist can be used together to provide a sustainability score or index. This is most useful when two similar developments, or variations of a single development, need to be compared. Scores should always be adjusted to reflect local conditions, and the particular emphasis of the development.

The principles underlying the issues: understanding the context

Good judgement in using the checklist will follow from a sound understanding of its underlying principles. 'Sustainability' is almost always context-dependent: what improves a problem in one area can exacerbate it in others, depending on the existing circumstances. For example, prosperity in one area may be threatened by high unemployment. Such an area may, however, have no shortage of affordable homes. Another area may have a severe problem filling available jobs as homes are expensive and in short supply.

It also follows that a fundamental rule of sustainable construction is that there must be an identified need for the development in the first place.

Also, although the built environment has a key role to play in sustainability, there are many other important unrelated factors, eg natural features, natural resources (water, arable land, mineral wealth, etc.), weather patterns, local culture, etc.

This checklist concerns itself with the underlying concerns and principles that relate to sustainability in the built environment. It addresses the inter-linked facets of sustainability: environmental, social and economic concerns and principles, described below.

Environmental concerns and principles

Environmental sustainability (in relation to the built environment) involves consideration of the local natural environment, and the national and global impacts. The main aspects of these are detailed below.

- Use land wisely and protect areas of natural beauty, scientific interest, etc.
- Use less energy and find more environmentally friendly forms of energy.
- Limit the amount of water treated for human consumption and increase the use of environmentally friendly water supply and drainage systems.
- Reduce the amount of road traffic to alleviate congestion, reduce air pollution and limit the land required for roads/car parks.
- Reduce the amount of raw materials used for construction, and consider appropriate means of extraction for materials that are plentiful.
- Encourage local sourcing of materials.
- Provide safe disposal of used materials that cannot be recycled.
- Protect and enhance wildlife and bio-diversity.

Social concerns and principles

The built environment should be a healthy, attractive and desirable place for people to live. Agreeing on the features to achieve this is not easy, but is likely to include the following aspects.

- A high 'quality' built environment (ie one that the majority of people find attractive and comfortable).
- A mix of housing (types and tenures).
- A mix of land uses (housing, employment and leisure).
- An appropriate density of buildings for the type of area.
- Provision of facilities/local centre, school, shop, chemist, etc.
- High accessibility throughout the area, with good public transport and provision for walkers and cyclists.
- A reduction in the domination of the car, particularly in residential areas.
- Measures to improve air quality.
- Provision of a high standard of urban design with sufficient public green space and areas of beauty.
- Designs to reduce the opportunities for crime.
- Designs to reduce noise nuisance and provide some quiet spaces.

Introduction

Economic concerns and principles

The requirements for economic sustainability vary depending on the nature of the community. Two extremes may be the densely urban and the isolated rural situations. The economic health of individual communities is linked to the economic health of the surrounding region, and it is the responsibility of the Regional Development Agencies (RDAs) to foster the economic growth within their regions. This checklist addresses the urban situation, and highlights some of the factors that make an area likely to be economically viable for the future. These are listed as follows.

- Providing employment sites to meet a projected need.
- Providing an appropriate intensity of land use to ensure viability of local business.
- Providing good infrastructure links to key trading centres by both public and private transport.
- Providing the basis for a thriving local economy.
- Supporting local trades and businesses during construction/regeneration activity.

Completing the checklist questions

These notes should be referred to whilst completing the checklist questions. They are designed to help the user understand the questions, the ranges of performance and the suggested scoring values.

1 Land use, urban form and design

Development land in the UK continues to be in short supply, and pressure remains high to provide the majority of new development on brown field sites or sites with low ecological value. In general development should be proposed in line with:
- *the land use guidance contained in the development plan; and*
- *current planning policy guidance notes and DETR Good Practice Guides.*

Guidance to planning authorities from DETR currently suggests that:

'Development which attracts a lot of people should be concentrated in or on the edge of existing towns or suburban centres, or be within areas which are or can be well served by public transport. Higher density housing should be encouraged within easy walking distance of these centres.'

Some brown field sites have a high ecological value and/or provide an open space (valued by residents), in a densely built-up urban environment. The 'value' of the brown field site should therefore be considered in any decision to redevelop.

Once an appropriate site has been chosen it is important to ensure that that the development is making the best use of the available site, and that urban spaces (in particular) are well designed. Creating quality urban spaces that people enjoy living and working in and which will stand the test of time, requires considerable skill. Some pointers towards 'good' urban design are provided in the DETR Good Practice Guide *By design*. Some of the key aspects from this guide like grain, layout, scale, building detail, and choice of materials, have been included in this checklist as far as possible.

Useful contacts

CIRIA
6 Storey's Gate, London SW1P 3AU

Commission for Architecture and the Built Environment (CABE)
7 St James's Square, London SW1Y 4JU

Department of Transport, Local Government and the Regions,
Eland House, Bressenden Place
London SW1E 5DU

English Partnerships
110 Buckingham Palace Road, London SW1W 9SB

Royal Town Planning Institute
26 Portland Place, London W1B 1LX

The Prince's Foundation,
19–22 Charlotte Road,
London EC2A 3SG

References

BRE. *Quiet homes*. BR 358. Garston, CRC, 1998.
Commission for Architecture and The Built Environment. *By design*. London, The Stationery Office. 2000.
Department of the Environment, Transport and the Regions (DETR). *Planning Policy Guidance Notes*. London, The Stationery Office.
    1 *General policy and principles*. 1997.
    3 *Housing*. 2000.
    6 *Town centres and retail developments*. 1996.
   12 *Development plans*. Revised edition. 1997.
   13 *Transport*. 2001.
   23 *Planning and pollution control*. 1994.
   25 *Development and food risk*. Draft in preparation, 2001.
DETR. *By design. Urban design in the planning system: towards better practice*. London, The Stationery Office. 2000.
English Partnerships. *The urban design compendium*. London, English Partnerships. 2000.
LPAC. *Open space planning in London*. London, LPAC. 1992.
The Prince's Foundation. *Sustainable urban extensions: planned through design*. London, The Prince's Foundation. 2000.
Urban Villages Forum. *Making places. A guide to good practice in undertaking mixed development schemes*. London, English Partnerships.

1.1 Site criteria

Objective
- To encourage the use of the most appropriate sites for development.

Questions

(a) Does this site meet the requirements of the Development Plan or other strategic planning guidance?

(b) Is the site free from planning constraints, eg
Designated flood plain?
Conservation Areas, including:
— ancient monuments and buildings?
— ancient landscapes?
— parks and gardens?
— SSIs (Sites of Special Scientific Interest)?
— AONBs (Areas of Outstanding Natural Beauty)?
Landfill sites?
Mineral extraction sites?

Answering the Questions

(a) **Site specific approval**
Check that this particular site meets a defined need and the policy objectives of the Development Plan, relevant *Planning Policy Guidance Notes* or other strategic independent report(s). (Other reports that identify need might include reports by the local Regional Development Agency on the economic needs of the area, reports on homelessness and unemployment, etc. by a recognised expert).
For example, *Planning Policy Guidance Note* 3 (Paragraph 31) contains a list of the preferred features for a site proposed for housing. However, these features will be different for different types of development, eg an industrial development.

(b) **Planning constraints**
Check that this particular site is free from any planning constraints. To achieve Best Practice, a site should be free of all constraints. To achieve Good Practice, the development must demonstrate how any constraints can be addressed such that they (a) have little effect on the proposed development, (b) cause no damage to existing valued features.

1.1 Site criteria

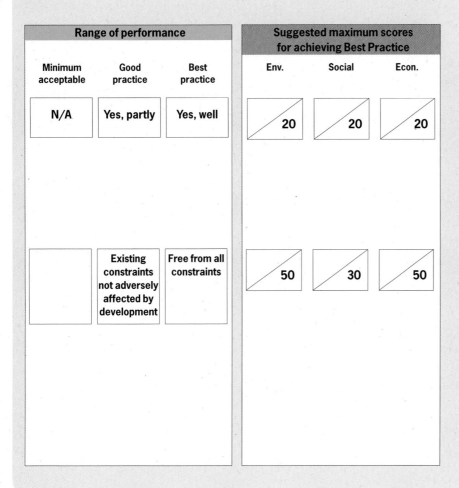

Range of performance			Suggested maximum scores for achieving Best Practice		
Minimum acceptable	Good practice	Best practice	Env.	Social	Econ.
N/A	Yes, partly	Yes, well	20	20	20
	Existing constraints not adversely affected by development	Free from all constraints	50	30	50

1.2 Reusing sites

Objective
- To save land and infrastructure resources.

Whilst it will generally be preferable to build on brown field sites near urban centres, it is recognised that the availability of such sites varies greatly within the UK. DETR guidance suggest that decisions about land use should be part of a strategic vision that looks ahead 25 years and considers what will be inherited by the next generation. The same guidance also suggests assessing the potential of an area by identifying possible development sites with good public transport links and existing facilities. In general, proposals should match with the reuse requirements of the Development Plan.

Questions	Answering the Questions
(a) Does the development reclaim any contaminated land?	**(a) Use of reclaimed/contaminated land** The checklist question asks about the % of contaminated land (expressed as a % of the total area) reclaimed by the development. The scoring system has been devised to reward the reclamation of contaminated land. However, the user may wish to adjust the scoring to reflect the level of contamination. *For example, to reclaim a small area of highly contaminated land may be as valuable as the reclamation of a large area of very low level contamination.* When dealing with contaminated land, the planning recommendations set out in DETR Circulars 17/89 for England and 38/89 for Wales and *Planning Policy Guidance Note 23* should be followed.
(b) Is the land decontamination method a sustainable option, ie *not* 'dig and dump' or 'cover layer' (see *section 1.3(c)* in these notes)?	**(b) Method of decontamination** There are many methods of remediating contaminated land and some of these have significant environmental consequences. The Construction Industry Research and Information Association (CIRIA) has produced 12 technical guidance documents on the remedial treatment of contaminated land. Two of the most common methods are given below. ● *Dig and dump* which means digging out the affected soil and removing the contaminated part to a land fill site elsewhere. Whilst this clears the site of the problem, it effectively moves the problem elsewhere. ● *Cover layer* describes the practice of placing a layer of inert ground (often clay) over the top of the contaminated land to seal it in. One of the concerns about this method is that over time the contaminants can leach out into water courses (unless the process is very carefully executed and monitored). As concerns over these two methods exist, in order to qualify for the additional points under this issue an alternative method must be used. Such methods include: ● biological decontaminant, ● soil vapour extraction, ● soil wash. There are also some negative environmental impacts associated with these methods, but they are generally of a lower order than the impacts associated with either 'dig and dump' or 'cover layer'.

1.2 Reusing sites

When considering developing brown field sites, thought should be given to the impact on the surrounding urban environment. Developments that improve the quality of the urban environment and provide amenity spaces in existing urban centres are important on many brown field sites. Whether a site is green field or brown field, it is important to establish its ecological value.

Many brown field sites contain some degree of contamination. The use of contaminated land is to be encouraged provided the proper processes of decontamination are followed. The method of decontamination can have a significant environmental impact, and in sustainability terms the method of decontamination is as important as the fact that the land is being reused.

Range of performance			Suggested maximum scores for achieving Best Practice		
Minimum acceptable	Good practice	Best practice	Env.	Social	Econ.
Multiply % (by area) of contaminated land by 0.25			25	25	25
Multiply % (by area) of contaminated land remediated (using a method other than 'dig and dump' or 'cover layer') by 0.5			50	50	50

Cont'd......

1.2 Reusing sites (cont'd)

Objective
- To save land and infrastructure resources.

Questions	Answering the Questions
(c) Does the development use any brown field land?	**(c) Use of brown field land** The Government has stated its intention that brown field sites be developed in preference to green field wherever possible. This checklist question rewards the use of brown field land. For the purposes of this checklist, contaminated land also counts as brown field (even if the contaminant is natural), and therefore the total area of brown field can be taken as the sum of the area of brown field plus the area of contaminated land.
(d) Does the proposal involve the release of any brown field land for redevelopment?	**(d) Brown field land released for redevelopment** If, as a result of the development proposal, brown field land will be released for redevelopment (either as part of the same site or in another area) credit is given. This can happen when an existing building is demolished and replaced by a building with a smaller footprint. The additional brown field land can then be made available for other uses. Another example of this is when a building is relocated on a new site (may even be a green field site) and the redundant brown field site is freed up for redevelopment.

1.2 Reusing sites (cont'd)

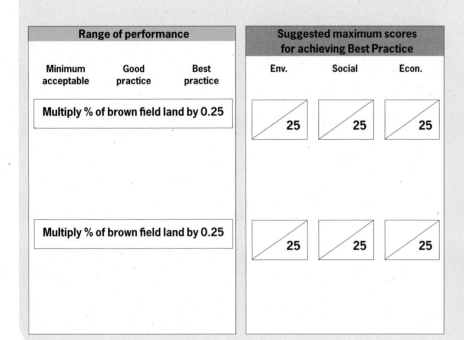

Range of performance			Suggested maximum scores for achieving Best Practice		
Minimum acceptable	Good practice	Best practice	Env.	Social	Econ.
Multiply % of brown field land by 0.25			25	25	25
Multiply % of brown field land by 0.25			25	25	25

1.3 Form of development: grain, layout, scale

Objective
- To ensure that the form of the development is appropriate for its sustained use.

The form a development takes affects many aspects of sustainability. The layout of streets and pavements and the siting of buildings determines the movement patterns within an area. Grain is a term used to describe the pattern made by the buildings and streets forming a development. Fine grain is used to describe small blocks with frequent divisions. The grain and scale of the development affects whether it is to be used by pedestrians or vehicles, whether it is suitable for domestic, commercial or industrial use, and the nature of the urban environment created.

Questions	Answering the Questions
(a) Is the grain of the development appropriate for needs, and in context with the surroundings?	**(a) Grain** If the grain of the development is both appropriate for the needs (eg human fine grain) and in context with the surroundings, Best Practice has been achieved. If the grain is well matched for needs but not so good for context, Good Practice has been met.
(b) Does the layout of the connecting roads, pavements, and spaces achieve a balance between good access into and through the development, and the provision of interesting and useful spaces?	**(b) Layout** The layout of the development in terms of its routes, paths and spaces is best when designed to be conducive to the provision of good access and the creation of interesting places. If it is good in both respects, Best Practice has been achieved. If it is good in one respect and fair in another, Good Practice has been met.
(c) Is the scale of development proposed appropriate in terms of the height and massing of the buildings?	**(c) Scale** The scale of urban development is described in DETR Good Practice Guide *By design* as a combination of the height of the development and its massing. If this fits well with the surroundings, the Best practice standard has been met. If it is good in one respect and fair in another, Good Practice has been met.

1.3 Form of development: grain, layout, scale

It is important that any new urban environment relates positively to the existing built environment. Good urban design can ensure that places are enjoyed and admired and remain vibrant and desirable.

Range of performance			Suggested maximum scores for achieving Best Practice		
Minimum acceptable	Good practice	Best practice	Env.	Social	Econ.
OK for needs and context	Well matched for needs, OK for context	Well matched for both needs and context	N/A	20	20
OK in both respects	Good in one respect and OK in the other	Good in both respects	N/A	20	20
OK in both respects	Good in one respect and OK in the other	Good in both respects	20	20	N/A

1.4 Open space/Landscaping

Objectives
- To ensure that the users of any new development have access to open public space.
- To ensure that landscaping opportunities are taken.

Questions

(a) Have quality green space and landscaping features been provided throughout the development, including boundaries?

(b) What % of the homes have access to public green space within 400 m of their door?

Answering the Questions

(a) Provision of green space/landscaping
If the opportunity to provide quality green space and landscaping features has been taken throughout the site, including boundaries, Best Practice has been achieved. If some aspects of landscaping have been provided but some opportunities have been missed, then Good Practice has been met.

(b) Access to green space
At least 50–60% of houses in the proposed development should have access to a play space/amenity space within 400 m of their front door to meet Good Practice. Best Practice is achieved if more than 60% of the houses in the proposed development (and over 30% of the other buildings) have access to an amenity space within 400 m of their front door.

1.5 Density

Objective
- To encourage high density development, where appropriate, in order to save land and ensure viability of, and accessibility to, local facilities.

Research carried out by McLoughlin for DETR showed that the greatest land savings occur by raising densities from 20 to 40 dwellings per hectare (DPH). As density increases to more than 40 DPH, it becomes economically viable to provide facilities such as shops within walking distance of the houses.

Planning Policy Guidance Note 3 clearly links density with accessibility and recommends high densities around major public transport nodes (up to 100 DPH). Less accessible areas should be used for low density development. As densities increase it is important that build quality remains high, especially sound insulation between dwellings (BRE 1998).

Questions

(a) What is the dwelling density?

(b) Has the density of the built environment been linked to public transport as recommended in *Planning Policy Guidance Note 13*?

Answering the Questions

(a) Dwelling density
The Good Practice standard recommends that at least 80% of the housing development (in terms of numbers of dwellings) is of greater density than 40 DPH (or equivalent habitable rooms per hectare [HRH]), and high standards of noise attenuation between dwellings have been adopted (BRE 1998). Best Practice will be achieved if 80% of the development is greater than 60 DPH (or equivalent HRH), with the same condition on noise attenuation.

(b) Density of all buildings related to public transport
The development proposal should follow a clear strategy of increasing development densities in line with access to public transport as outlined in *Planning Policy Guidance Note 13*. If this is fully demonstrated, Best Practice has been achieved. If only partly demonstrated, Good Practice has been achieved.

1.4 Open space/Landscaping

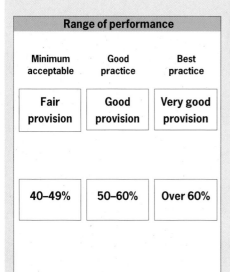

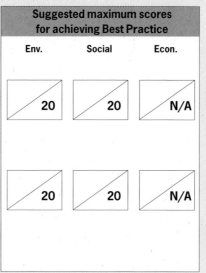

1.5 Density

Also, high density residential developments need good access to recreation and leisure facilities. DETR guidance suggests that it is also possible to increase the density of non-residential development provided the public transport system is in place to support this.

Range of performance			Suggested maximum scores for achieving Best Practice		
Minimum acceptable	Good Practice	Best Practice	Env.	Social	Econ.
35–40 DPH	41–59 DPH	> 60 DPH	20	20	20
	Yes, partly	Yes, well	20	20	20

1.6 Mix of uses

Objective
- To encourage a variety of building types (including homes) to be built in close proximity.

Mixed use development encourages people to work and live close by, and reduces the requirement for motorised transport. It has other advantages from a sustainability point of view, namely:
- *It encourages areas to remain occupied for most of the day, reducing the crime risk.*
- *It provides opportunities to include local forms of power supply, eg renewables and combined heat and power (CHP) which are often more environmentally benign than national sources. CHP is most efficient when there is a constant demand for the heat that is produced with the electricity. Hospitals, hotels and industry often need heat all year and make ideal sites for combined heat and power generation. By grouping other developments (like dwellings), around main CHP plants they can also be provided with environmentally friendly heat and power (see chapter 3, Energy).*

Questions

(a) Does the mix of uses in the proposed development meet the requirements of the Development Plan?

(b) Does the % of affordable homes provided meet the requirements of the Development Plan and housing needs surveys?

(c) Has the affordable housing been sensitively integrated with the rest of the development in terms of aesthetics and distribution?

(d) Has a retail impact study been carried out?

(e) What is the percentage of flexible buildings (out of the total number of buildings)?

Answering the Questions

(a) Mixed use
Best Practice is achieved by mixed use development that fully meets the requirements of the Development Plan. If some opportunities for mixed use have been taken but others missed, Good Practice has been met.

(b) Meeting the target on affordable homes
Good Practice has been met if the % of affordable housing is close to (but less than) the requirements of the Development Plan/housing needs study. If the % of affordable homes meets or exceeds the targets in the Development Plan, Best Practice has been met.

(c) Integration of affordable homes
Is the provision of affordable homes well integrated (ie interspersed) into the rest of the development, and are the homes aesthetically in keeping with the rest of the development? Best Practice has been achieved if both have been met.

(d) Retail impact study
Planning Policy Guidance Note 6 recommends a retail impact study as part of any proposal to build significant new retail premises, especially those designated as 'out of town'. If a study has been carried out and the existing retail premises are likely to be largely unaffected by the new development, then Good Practice has been met. If existing retail premises are likely to have enhanced viability, then Best Practice has been met.

(e) Flexible buildings
Flexibility in buildings takes a number of forms:
- Buildings that can be converted from one use to another (offices into flats, large homes into smaller units, homes into workplaces, etc.).
- Buildings that can be easily extended or modified. This could include homes with roofs and basements that can be converted into living space – or factory produced 'modular' buildings that can have modules added or taken away without affecting the structure of the remaining building.
- Buildings that can easily be relocated should the need for that building change.

If between 10% and 25% (in number of buildings) have any of the kinds of flexibility described above, Good Practice has been achieved. If over 25% of buildings have been designed to be flexible, Best Practice has been achieved.

1 Land use, urban form and design

1.6 Mix of uses

The appropriate mix of development will depend on the type of area involved, ie city, town centre, neighbourhood, urban village, etc. It is desirable to have a mixture of domestic, employment and social facilities within an urban area, and to provide a range of housing types including affordable housing. This enables a full mix of different social groups in the same neighbourhood, and can contribute to social integration. If, in addition, a proportion of buildings are designed to be flexible in their use, changes in economic circumstances can be accommodated without the need for redevelopment. This can ensure that the buildings have long useful lives, even if the purpose for which they were initially built changes.

Range of performance			Suggested maximum scores for achieving Best Practice		
Minimum acceptable	Good practice	Best practice	Env.	Social	Econ.
	Yes, partly	Yes, well	10	10	10
	Yes, within a few %	Well above requirement	N/A	10	10
	Yes, to one aspect	Yes, to both aspects	N/A	10	10
	Yes, largely unaffected	Yes, enhanced	10	10	10
	10–25%	> 25%	10	10	10

1.7 Aesthetic aspects

Objective
- To encourage attractive high quality development details and materials.

The aesthetic aspects of a development proposal are notoriously difficult to assess as 'attractiveness' is subjective. However, panels judging the quality of development proposals will often agree on which design is the most 'attractive' without necessarily being able to specify why.

Questions

(a) Is the appearance of the development, in relation to the detailed building elements (eg roofscapes, window details, etc.), both attractive and in context?

(b) Is the appearance of the development (in terms of the choice of building materials, ie colour, form, variety and durability), both attractive and in context?

Answering the Questions

(a) Building details
Examine the buildings details (windows, roof details, etc.) to see if they are both attractive and in context with the local surroundings. If the details are carefully considered and meet both requirements well, Best Practice has been achieved. If the details are better in one aspect than another, Good Practice has been met.

(b) Choice of materials
Examine the choice of building materials. Are the materials chosen both attractive and in context with the local surroundings(in terms of their colours, texture, variety and durability)? In addition, can the majority of the materials chosen be obtained locally? If the materials are attractive, in context and available locally, Best Practice has been met. If the materials are attractive and in context but cannot be obtained locally, Good Practice has been met.

1 Land use, urban form and design

1.7 Aesthetic aspects

The checklist concentrates on two specific aspects of a development, building details and materials, as essential components of the detailed urban design. Without attention to details like these many good designs fail to deliver a lasting attractive environment.

Range of performance			Suggested maximum scores for achieving Best Practice		
Minimum acceptable	Good Practice	Best Practice	Env.	Social	Econ.
	Carefully considered	Very carefully considered	20	20	N/A
	Attractive and in context	Attractive, in context and local	20	20	N/A

Total for Land use, urban form and design | 365 | 405 | 325

2 Transport

Transport is responsible for environmental, social and economic impacts. Locally, it results in noise, air and water pollution, and congestion, and it can either prevent or provide access. Globally, transport is a major user of fossil fuel and contributes significantly to global warming.

Lack of access has significant social implications too, by isolating certain sections of society and lowering their quality of life. Poor access also affects business prosperity.

The way that we position buildings has a strong influence on how fully they will be used, and the amount of energy used to transport people between them. In general, planning guidance advises local authorities to encourage development in areas that are well served by public transport. Where new areas are proposed, regional and local plans should focus development around public transport nodes and corridors. Residential development density should be kept high around railway stations, etc. For new settlements, such as 'urban villages', it is important to have good public transport links to existing major employment centres as even with mixed use development many new residents may not be able (or willing) to work locally.

The checklist assumes that facilities must be within 500 m walking distance for people to walk or cycle to them regularly. In addition, pedestrian footpaths and cycle lanes linking the key parts of the development must be provided if environmentally friendly forms of transport are to be encouraged. In exceptional cases, it may be appropriate to give credit to facilities that are further away than 1 km if there is frequent cheap public transport. These assumptions are supported by a number of studies including *The urban design compendium*.

Consideration should be given to the financial viability of operating a public transport service. By placing major travel generators (like hospitals) at either end of the transport corridor, the two-way flow of traffic along a corridor can be increased, making a public transport service across the whole length of the corridor sustainable. To serve a new development successfully, public transport provision must be available from the beginning of the development's life. (*Note:* this will require subsidies during the early period.) Otherwise, car use patterns will become established and will be difficult to influence subsequently.

Accessibility is also important to encourage the development of a 'community'. The design of cycle ways and footpaths must consider personal safety as they will only be used if people feel safe.

Useful contacts	References

Department of Transport, Local Government and the Regions, Eland House, Bressenden Place London SW1E 5DU	**Department of the Environment, Transport and the Regions (DETR).** *Planning Policy Guidance Notes.* London, The Stationery Office.
	   3 *Housing.* 2000.
	   6 *Town centres and retail developments.* 1996.
	  13 *Transport.* 2001.
	Department of the Environment, Transport and the Regions. *Preparing your organisation for transport in the future: the benefits of green transport plans.* London, DETR. 1999.
	English Partnerships. *The urban design compendium.* London, English Partnerships. 2000.
	Llewelyn Davies & JMP Consultants. *Parking standards in the South East.* London, Government Office for the South East, DETR. 1998.

2.1 General policy

Objective
- To ensure that the transport plans proposed for the development agree with those within the local plan.

Many local authorities have transport policies/plans within the Local Plan. The purpose of the policy will generally be to maintain high levels of accessibility and encourage the use of environmentally friendly forms of transport, whilst reducing traffic congestion, air pollution and the need for travel.

Questions

(a) Does the development meet the requirements in the Local Transport Plan?

(b) Is the development within an existing transport corridor, growth point or node?

(c) Have travel surveys been carried out to research existing travel patterns and increase the understanding of travel needs?

(d) Has a Traffic Impact Assessment been carried out, and were the results beneficial?

Answering the Questions

(a) Match with transport policy
Check that the proposed development follows the principles and standards laid down in the transport policy. Best Practice is full implementation. Good Practice, and the minimum requirement, is partial implementation.

(b) Good transport links and existing transport corridor
Developments that are largely (more than 80% of the development footprint) within an existing transport corridor, growth point or node meet Best Practice. If over 30% (of the development footprint) lies within the corridor, Good Practice has been met.

(c) Travel surveys
Check whether any travel surveys are to be carried out to establish existing travel patterns in and around the area of the proposed development. If comprehensive travel surveys have been carried out, Best Practice has been achieved.

(d) Traffic Impact Assessment (TIA)
Has a traffic impact assessment including the following key features been carried out?
- Description of existing conditions.
- Consideration of the effects of the development on all road users and residents.
- Consideration of the effects of the development on the environment and the community.

If a traffic impact assessment covering all the above aspects has been carried out and shows either positive or very few negative impacts, Best Practice has been achieved. If the benefits highlighted by the TIA match the negative impacts identified, Good Practice has been achieved.

2.1 General policy

Range of performance			Suggested maximum scores for achieving Best Practice		
Minimum acceptable	Good practice	Best practice	Env.	Social	Econ.
N/A	Yes, partly	Yes, well	20	20	20
N/A	Yes, > 30%	Yes, > 80%	20	20	20
N/A	Yes, partly	Yes, well	20	20	20
None done	Benefits balance impacts	Benefits outweigh impacts	25	25	25

2.2 Public transport provision

Objectives
- To reduce road congestion and air pollution by reducing single occupancy private car journeys.
- To enable access by the whole community, ie not just car drivers.

Questions	Answering the Questions
(a) What is the distance from a major fixed public transport node (train, tube, tram), or regular link (every 10–15 mins) to a major fixed public transport node, for 50% of the footprint?	**(a) Proximity to fixed public transport node** If 50% of the footprint of the proposed development is within 2 km of a major fixed public transport node, Good Practice has been met. If 50% of the development footprint is within 1 km of a major fixed transport node, Best Practice has been met.
(b) What is the distance from the bus stop or other public transport node (new or existing), providing a regular service. (80 % of the development to fall within this)?	**(b) Convenience of public transport** If 80% of the footprint of the proposed development is within 1 km of a public transport stop that receives a regular service, Good Practice has been met. If 80% of the footprint of the proposed development is within 500 m of a transport stop that receives a regular service, Best Practice has been met. *Note:* a regular service is defined as having a service every 20 minutes or less at peak times, and every 45 minutes for the rest of the day.
(c) Has provision been made for a comfortable/safe bus shelter or waiting room near local activities? [6]	**(c) Design and location of bus shelters** If over 80% of the proposed development is served by convenient bus shelters, Best Practice has been met. If 60–80% is served by convenient bus shelters, Good Practice has been met.
(d) What percentage of the bus stops and shelters have a real time information system?	**(d) Provision of real time information systems** If provision of real time information systems to the majority of the bus stops/bus shelters on the site (ie more than 60%) is proposed, Good Practice has been met. If > 80% of bus stops have real time information, Best Practice has been met.
(e) Has provision been made for environmentally friendly public transport (frequent service) to the city/town centre (bicycle, gas bus, cycle rickshaw, etc.)?	**(e) Provision of environmentally friendly public transport** Good Practice has been met if 10–25% of the development is served by environmentally friendly public transport (likely to emit virtually no CO_2). If over 25% of the development is served by environmentally friendly public transport, Best Practice has been met.

[6] Number and size of bus shelters should be in proportion to the final population.

2.2 Public transport provision

Range of performance			Suggested maximum scores for achieving Best Practice		
Minimum acceptable	Good practice	Best practice	Env.	Social	Econ.
	1–2 km	< 1 km	20	20	20
	≤ 500 m	≤ 200 m	10	10	10
	60–80%	> 80%	20	20	20
	60–80%	> 80%	10	10	10
	Serves 10–25% of development	Serves > 25% of development	10	10	10

2.3 Parking

Objective
- To control the number and cost of parking places available in order to reduce the number of car journeys.

Planning Policy Guidance Notes PPG 13 and PPG 6 advise local authorities to adopt a comprehensive strategy to cover car parking. PPG 3 suggests that the number of off-street car parking spaces on urban housing developments should be restricted to an average of 1.5 spaces per dwelling. Restricting the availability of parking spaces must be linked with the provision of adequate public transport, otherwise accessibility and economic viability will suffer.

Questions | Answering the Questions

(a) Have the transport needs of different users been provided for by balancing the availability/charging of car parking with adequate public transport?

(a) Needs of users
Does the proposal demonstrate that the transport need of the users of the development have been carefully considered and provided for? If the needs of the users have been thoroughly covered, Best Practice has been met. To demonstrate this a proposal would usually require the following:
- Identification of the relevant user groups.
- Provision of the infrastructure to provide appropriate transport options for each group.
- Provision of a balance of transport options to support the car parking option adopted.

If a proposal contains some of these attributes but not all, or is weak in any of these aspects, Good Practice has been met.

(b) How do the car parking standards for the development compare with the Local Authority requirements?

(b) Car parking standards
If the development has adopted car parking standards below the Local Authority maximum, but above the local authority minimum, Good Practice has been met. If the development has adopted car parking standards in line with the Local Authority minimum, Best Practice has been met. Practice).

(c) What % of car parks have been designed to be flexible?

(c) Flexible design of car parks
Flexible car parking is hard standing parking that has been designed to have more than one use, eg it can become an amenity space (hard play ground, market square, picnic area, etc.)? If 10–20% of the parking has been designed to have an alternative use, Good Practice has been met. If more than 20% has been designed to have an alternative use, Best Practice has been met.

(d) Will there be a reduction of the visual impact of parking by screening?

(d) Reduction of visual impact/attractive integration
Have efforts been made to reduce the visual impact of the car parking and to attractively integrate it into the development? With *Planning Policy Guidance Note 3* seeking to limit off-street parking for dwellings, imaginative design layouts are required to accommodate both the on-street and the off-street parking spaces. Methods to reduce the impact of the car park include screening (often by planting or earth banks) or situation (siting car parks behind or underneath buildings, or in courtyards). If 30–50% of the car parking has been designed to reduce the impact of the car park or to integrate well with the design, Good Practice has been met. If over 50% of the parking has been designed to reduce impact or to integrate well with the design, Best Practice has been met.

(e) Has provision been made for off-road HGV unloading spaces or alternatives?

(e) HGV offloading spaces
Where HGV access is necessary, has provision been made for these large vehicles to unload and turn round off the main route through the development? If this has been provided for more than 60% of the cases, Good Practice has been met. If this has been provided in 80% or more of cases, Best Practice has been met.

2 Transport

2.3 Parking

Range of performance			Suggested maximum scores for achieving Best Practice		
Minimum acceptable	Good practice	Best practice	Env.	Social	Econ.
	Partly considered or provided	Fully considered or provided	30	30	30
	< LA max. = 5	Equal to LA min. = 10	10	10	10
	10–20%	> 20%	10	10	10
	In 30–50%	Yes, in > 50%	10	10	N/A
	Yes, for > 60% of cases	Yes, for > 80% of cases	10	10	10

2.4 Facilities for pedestrians and cyclists

Objectives
- To provide a safe and welcoming environment for pedestrians and cyclists and so encourage residents to walk and cycle for most of their local journeys.
- To make the development accessible to all groups in the community.

Questions	Answering the Questions
(a) Will there be a network of safe pavements around site and to local facilities?	(a) **Provision of safe pedestrian routes** Check that the development provides a network of safe pedestrian routes around the development and to local facilities. If full provision is made, Best Practice has been met.
(b) Has provision been made for safe crossing points over all major roads near to facilities and at strategic points?	(b) **Provision of safe crossing points** Check that the planning proposal contains the provision of safe crossing points over all major roads near to facilities and at strategic points. If full provision is made, Best Practice has been met. If only partial provision is made, Good Practice has been met.
(c) Is there a network of safe bicycle routes[7] to local facilities near to, and overlooked by, roads and pavements?	(c) **Provision of cycle routes** Check that the proposal contains a network of safe cycle routes around the development. If fully provided, Best Practice has been met. If only partial provision is made, Good Practice has been met.
(d) Has provision been made for safe bicycle storage at all local facilities and at strategic points?	(d) **Provision of cycle storage** Check that the proposal contains safe bicycle storage places at local facilities and strategic points. Partial provision is Good Practice. Full provision is Best Practice.

[7] Route can be shared.

2.5 Provision of local employment

Objective
- To enable people to live and work in close proximity, limiting the need to travel to work by car.

Questions	Answering the Questions
(a) For a development providing significant numbers of housing, what is the ratio (in %) of land occupied by easily accessible (within 1 km radius) employment sites:housing?	**Proximity of local employment sites to housing** Consider the area of housing in the proposed development and calculate approximately the area of employment sites (accessible by foot/cycle) within a 1 km radius of the centre of the housing development. If the area of employment is between 15 and 30%, Good Practice has been met. If the area of employment is more than 30% of the area of housing, Best Practice has been met.
(b) For a development which is predominantly commercial in its use, what is the ratio (in %) of land occupied by easily accessible housing:employment sites?	**Proximity of housing to commercial development** Consider the area of commerical buildings in the proposed development and calculate approximately the area of housing within easy access of the commercial development (ie within a 1 km radius of the centre of the commercial development). If the area of housing is more than 40%, Good Practice has been met. If the area of housing is more than 60%, Best Practice has been met.

2.4 Facilities for pedestrians and cyclists

Range of performance			Suggested maximum scores for achieving Best Practice		
Minimum acceptable	Good practice	Best practice	Env.	Social	Econ.
		Full provision	10	10	10
	Partial only	Full provision	10	10	10
	Partial only	Full provision	10	10	10
	Part provision	Full provision	10	10	10

2.5 Provision of local employment

Range of performance			Suggested maximum scores for achieving Best Practice		
Minimum acceptable	Good practice	Best practice	Env.	Social	Econ.
	25–30%	31% or more	10	10	10
	> 40%	> 60%	10	10	10

2.6 Proximity of local facilities, ie within easy walking distance (500 m) of > 60% of homes and offices

This issue concerns the proximity of local services to housing and commercial development

Objective
- To reduce the need for residents and office staff to travel by car to essential facilities thereby reducing the associated pollution and congestion.
- To help build a community focus and ensure social equity by enabling all in society to have access to essential facilities.

Questions

Which of the following are within 500 m of the development:

Answering the Questions

- **Proximity of local facilities to homes**
  The recommendation has been met for each of the facilities (a)–(m) below, that are within walking distance (500 m) of 60% (by number of dwellings) of the housing.
- **Proximity of local facilities to commercial development**
  The recommendation has been met if the facilities a, c, e, f, g, i, l, and m (facilities b, d, h, j, k are not relevant to offices) are within walking distance of 60% (by number) of the offices/commercial premises.

(a) Shop selling food including fresh groceries?

(b) School?   JMI = 10,
                Secondary = 20

(c) Playground/amenity area?

(d) Local meeting place?

(e) Medical centre?

(f) Chemist (Pharmacy)?

(g) Leisure facilities including public house?

(h) Childcare facilities (nursery/crèche)?

(i) Post box/phone box?

(j) Religious building/place of worship?

(k) Cemetery?

(l) Contemplative features (water garden, etc.)?

(m) Cash-point machine?

2.6 Proximity of local facilities, ie within easy walking distance (500 m) of > 60% of homes and offices

Range of performance			Suggested maximum scores for achieving Best Practice for domestic and commercial development		
	Homes	Offices	Env.	Social	Econ.
(a) Shop selling food inc. fresh groceries	Yes (10)	Yes (10)	20	20	N/A
(b) School: JMI = 10, Secondary = 20	10 or 20	N/A	30	30	N/A
(c) Playground/amenity area	Yes (10)	Yes (10)	20	20	N/A
(d) Local meeting place	Yes (10)	N/A	10	10	N/A
(e) Medical centre	Yes (10)	Yes (10)	20	20	N/A
(f) Chemist (Pharmacy)	Yes (10)	Yes (10)	20	20	N/A
(g) Leisure facilities, inc. public house	Yes (10)	Yes (10)	20	20	N/A
(h) Childcare facilities, eg creche/nursery	Yes (10)	Yes (10)	20	20	N/A
(i) Post box/Phone box	Yes (10)	Yes (10)	20	20	N/A
(j) Religious building/place of worship	Yes (10)	N/A	10	10	N/A
(k) Cemetery	Yes (10)	N/A	10	10	N/A
(l) Contemplative features	Yes (10)	Yes (10)	20	20	N/A
(m) Cashpoint machine	Yes (10)	Yes (10)	20	20	N/A

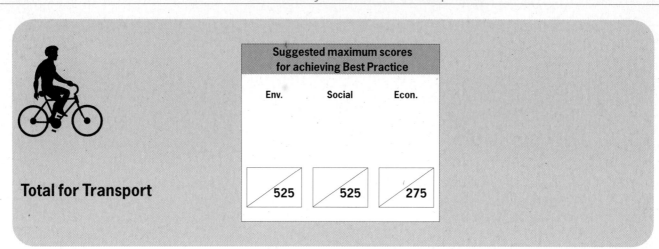

Total for Transport

3 Energy

An increase in the build-up of carbon dioxide in the earth's atmosphere caused by the burning of fossil fuels is causing global warming. This in turn is changing the world's weather (climate change). Many environmental specialists consider the burning of fossil fuels to be the most serious environmental concern currently facing the world. The full effects of global warming are not known but are likely to include:
- *warming of the oceans (causing the destruction of coral, etc.),*
- *melting of the polar ice caps,*
- *flooding due to changes in the levels of oceans,*
- *increased severe storms and weather patterns causing crop failure, drought, flooding and fire.*

Servicing buildings (including housing) accounts for approximately 50% of the UK's total energy and CO_2 consumption. The need to reduce the consumption of fossil fuels must be central to any policy on sustainability. In terms of construction, energy is used in the following ways:
- to manufacture and transport building materials and products ('embodied energy'),
- to create the structure on site,
- to operate the building/structure throughout its lifetime ('in-use energy'),
- to transport people and goods between the buildings (transport energy),
- to demolish and transport demolition waste.

Energy from renewable sources (solar, hydro, wind, geothermal, etc.) does not add to climate change and has the benefit of being infinitely available. This chapter considers the energy provision for the whole development rather than individual buildings (which is covered in Chapter 4 *Impact of individual buildings*). At the site/development level, many of the renewable/alternative energy sources become more viable.

Useful contacts

BRECSU manages the UK Government's Energy Efficiency Best Practice programme (EEBPp). Free copies of EEBPp guides can be obtained by telephoning 0800 585749.

Climate Care Label of The Carbon Trust, www.CO2.org.

Combined Heat & Power Association (CHPA), 35–37 Grosvenore Gardens, London SW1W 0BS.

Energy Saving Trust, www.energysavingtrust.co.uk.

Energy Technology Support Unit (ETSU). Energy Efficiency Enquiries Bureau, ETSU, Harwell, Oxfordshire OX11 0RA (Tel 01235 436747).

References

Department of the Environment, Transport and the Regions. *Renewable energy. Planning Policy Guidance Note 22.* London, The Stationery Office. 1994.

Department of the Environment, Transport and the Regions. *Introduction to large-scale combined heat and power.* Good Practice Guide 43. Harwell, Oxfordshire, ETSU. 1999.

Department of the Environment, Transport and the Regions. *Guide to community heating and CHP.* Good Practice Guide 234. London, Department of the Environment, Transport and the Regions. 1998.

3.1 Community-wide energy production

Objective
- To reduce carbon dioxide emissions from the new development.

A number of opportunities to provide environmentally friendly energy exist when considering a community wide development. Renewable schemes like hydro and wind generation have a critical mass for operation that is usually larger than the energy requirements of a single building. Similarly, Combined Heat and Power (CHP) generation becomes viable for a whole community, especially one in which there is mixed use development.

Questions

(a) Has the site been assessed for its suitability for renewable energy production?

(b) What % of energy is produced from a community-wide renewables scheme (eg wind farm, hydro scheme, photovoltaic bank, CHP operating on biomass or waste)?

(c) What % of energy is produced from a CHP unit (running on fossil fuel), as a proportion of the total energy requirement?

Answering the Questions

(a) Assessing the potential for renewable energy
Consideration of the availability, practicality and cost of including renewable sources should be included in any large-scale development proposal. If there has been a full assessment of the potential, Best Practice has been met. If a limited assessment has been performed, Good Practice has been met. *Note:* reference should be made to *Planning Policy Guidance Note 22*.

(b) Contribution from renewable energy
An estimate of the total amount of energy (in % of total) supplied to the development from a renewable source should be calculated and multiplied by the appropriate factor (see table, right) to give the points score in all three categories of sustainability.

(c) Combined heat and power (CHP)
An estimate of the amount of energy (as a % of the total) supplied to the development from CHP plant fuelled by fossil fuel should be entered here. Multiply this by the appropriate factor (see table, right) to give the points score in all three categories of sustainability.

3 Energy

3.1 Community-wide energy production

CHP is efficient because it utilises the heat (generated in electricity production) that is usually wasted at the power station. Typically, the conversion factor from fuel to power at a large power station is only 25%! CHP units use a variety of different fuel sources, including premium grade fuel like natural gas, commercial grade (eg heavy oils and coal), and waste (biomass, domestic waste). CHP which runs from biomass (including wood) and waste is considered to be a renewable supply. Currently, grants are available for feasibility studies to be carried out on CHP. Such studies are essential as the financing, technical and maintenance aspects of CHP require detailed consideration by experts. Information on CHP is available free from BRECSU and ETSU (see Useful contacts and References).

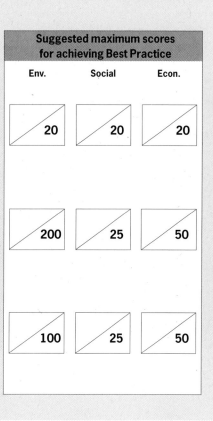

Range of performance			Suggested maximum scores for achieving Best Practice		
Minimum acceptable	Good practice	Best practice	Env.	Social	Econ.
	Partly assessed	Fully assessed	20	20	20
Multiply % of renewable energy by the following factors: Env. = 2.0, Social = 0.25, Econ. = 0.5			200	25	50
Multiply % by the following factors: Env. = 1.0, Social = 0.25, Econ. = 0.5			100	25	50

3.2 Street lighting

Objective
- To ensure that street lighting is as energy efficient as possible, and minimises upward light pollution.

Questions

(a) Is there energy-efficient street lighting with limited upward light transmission?

Answering the Questions

(a) Energy-efficient street lighting
Assess the proportion of street lights that are low energy (high pressure sodium high intensity discharge lamps, ie SON) or run off renewable energy. If the proportion of lights is 70–80%, Good Practice has been met. If the proportion of lights is over 80%, Best Practice has been met.

3.2 Street lighting

Range of performance			Suggested maximum scores for achieving Best Practice		
Minimum acceptable	Good practice	Best practice	Env.	Social	Econ.
	70–80%	> 80%	/20	/10	/10

Total for Energy /340 /80 /130

4 Impact of individual buildings

In order to build in a sustainable manner, it is necessary to minimise any negative impacts. The main impacts attributed to the construction of individual buildings are:
- *energy in use,*
- *embodied energy and main environmental impacts of building materials,*
- *water consumption,*
- *health and wellbeing of occupants: indoor air quality/daylighting/noise,*
- *transport and access impacts of occupants and users,*
- *pollution to air (CO_2, SO_X and NO_X), ozone depletion.*

The impacts of individual buildings can be addressed in two ways.
- Set a target using a proprietary quality-assured rating scheme like BREEAM (Building Research Establishment Environmental Assessment Method). The use of a rating system like BREEAM enables all the above issues to be addressed in one target; it also allows developers a degree of flexibility in the way that they meet the 'sustainability' target. For planning authorities and developers alike it gives an independent way of 'telling the time' so that all development standards can be judged on a level playing field.
- Specify targets for each aspect of sustainability. This can involve considerable effort on the part of those assessing the scheme, as rigid targets must be tailored for each separate development under consideration. Such targets offer developers much less flexibility in the combination of measures that they adopt to achieve a high standard of sustainability. General guidance on setting targets has been given as far as possible, but does not cover all issues for all building types.

Useful contacts	References
BRECSU manages the UK Government's Energy Efficiency Best Practice programme (EEBPp). Free copies of EEBPp guides can be obtained by telephoning 0800 585749. **BREEAM**, BRE, Bucknalls Lane, Watford WD25 9XX. www.bre.co.uk. Email: Breeam@bre.co.uk **CO_2 conversion factors** www.energy-efficiency.gov.uk/document/factfigs/emiss.htm **Energy Saving Trust** www.energysavingtrust.co.uk. **Recycled/reclaimed building materials** Materials Information Exchange at http://helios.bre.co.uk/waste **Sustainable timber** *Forests Forever* at www.forestsforever.co.uk, or UK Woodland Assurance Scheme (UKWAS), email: ukwas@forestry.gov.uk	**Anderson J & Howard H.** *The green guide to housing specification.* BR 390. Garston, CRC. 2000. **Anderson J & Shiers D.** *The green guide to specification.* 3rd edition. Oxford, Blackwell Scientific. 2002. **Baldwin R, Yates A, Howard N & Rao S.** *BREEAM '98 for offices.* BR 350. Garston, CRC. 1998. **BRE.** *BREEAM Version 2/91. An environmental assessment for new superstores and supermarkets.* BR 207. Garston, CRC. 1991. **BRE.** *BREEAM/New industrial units. Version 5/93. An environmental assessment for new industrial, warehousing and non-food retail units.* BR 252. Garston, CRC. 1993. **BRE.** *Quiet homes: a guide to good practice and reducing the risk of poor sound insulation between dwellings.* BR 358. Garston, CRC. 1998. **Hobbs G & Collins R.** *Demonstration of reuse and recycling of materials: BRE energy efficient office of the future.* BRE Information Paper IP5/94. Garston, CRC. 1994. **Howard N, Edwards S & Anderson J.** *BRE methodology for environmental profiles of construction materials, components and buildings.* BR 370. Garston, CRC. 1999. **Littlefair P.** *Site layout planning for daylight and sunlight: a guide to good practice.* BR 209. Garston, CRC. 1991. **Littlefair P J.** *Site layout for sunlight and solar gain.* BRE Information Paper IP4/92. Garston, CRC. 1992. **Littlefair P J.** *Site layout planning for daylight.* BRE Information Paper IP5/92. Garston, CRC. 1992. **Rao S, Yates A, Brownhill D & Howard N.** *EcoHomes: the environmental rating for homes.* BR 389. Garston, CRC. 2000.

4.1 Meeting a specified BREEAM rating

Objective
- To make a commitment to deliver a certain BREEAM rating standard for all relevant building types (ie homes, offices, factories and light industrial units and supermarkets).

The BREEAM scale runs from PASS through GOOD and VERY GOOD to EXCELLENT. PASS level is designed to ensure that the developer has considered and addressed the full range of environmental issues. To reach a GOOD standard should be possible with good knowledgeable design from inception. To reach a standard of VERY GOOD is likely to involve some additional capital cost. EXCELLENT is unlikely to be reached without some aspect of innovative design and servicing of a building. Some site issues like poor public transport may make it very difficult for a building to score an 'EXCELLENT' so the standards must be applied in a realistic way with some room for negotiation depending on the site conditions.

Questions

What is the BREEAM rating for the proposed buildings? For a building not covered by BREEAM, use individual targets (see section 4.2 below).

Answering the Questions

BREEAM rating
Enter % of buildings for each type expected to score a BREEAM rating of PASS, GOOD, VERY GOOD or EXCELLENT in boxes 1, 3 and 5. Multiply by the relevant scoring factor (SF) and fill in boxes 2, 4 and 6. Sum the values in boxes 2, 4, and 6 and enter in box 7. Multiply value in box 7 by the sustainability factor to give scores for Environmental, Social and Economic factors in boxes 8, 9 and 10.

(a) EcoHomes (the BREEAM version for homes)

Totals

(b) BREEAM for Offices

Totals

(c) BREEAM for Factories/Light industrial

Totals

4 Impact of individual buildings

4.1 Meeting a specified BREEAM rating

The development should have a BREEAM assessment carried out on each relevant building by a qualified assessor. The checklist suggests that PASS is the minimum acceptable standard. Higher points are awarded in the scoring system for meeting a higher BREEAM rating which is reflected in the scoring factor. To gain the final score in the three sustainability categories, divide by the number of different building types.

Range of performance			Suggested maximum scores for achieving Best Practice		
Minimum acceptable	Good practice	Best practice	Env.	Social	Econ.
PASS SF = 0.25	GOOD SF = 0.5	VERY GOOD SF = 0.75 EXCELLENT SF = 1.0	**4.0** for all building types	**2.0** for all building types	**2.0** for all building types
1: %	2: % × SF	3: %	4: % × SF	5: %	6: % × SF
			7: /100	8: /400	9: /200	10: /200
1	2	3	4	5	6
			7: /100	8: /400	9: /200	10: /200
1	2	3	4	5	6
			7: /100	8: /400	9: /200	10: /200

Cont'd.....

4.1 Meeting a specified BREEAM rating (cont'd)

Objective
- To make a commitment to deliver a certain BREEAM rating standard for all relevant building types (ie homes, offices, factories and light industrial units and supermarkets).

Questions

What is the BREEAM rating for the proposed buildings?
For a building not covered by BREEAM, use individual targets (see section 4.2 below).

Answering the Questions

BREEAM rating
Enter % of buildings for each type expected to score a BREEAM rating of PASS, GOOD, VERY GOOD or EXCELLENT in boxes 1, 3 and 5. Multiply by the relevant scoring factor (SF) and fill in boxes 2, 4 and 6. Sum the values in boxes 2, 4, and 6 and enter in box 7. Multiply value in box 7 by the sustainability factor to give scores for Environmental, Social and Economic factors in boxes 8, 9 and 10.

(d) BREEAM for Superstores and Supermarkets

Totals

4.1 Meeting a specified BREEAM rating (cont'd)

Range of performance

Minimum acceptable	Good practice	Best practice
PASS SF = 0.25	**GOOD** SF = 0.5	**VERY GOOD** SF = 0.75 **EXCELLENT** SF = 1.0
% (1)	% × SF (2)	% (3)	% × SF (4)	% (5)	% × SF (6)

Total: 7 / 100

Suggested maximum scores for achieving Best Practice

Env.	Social	Econ.
4.0 for all building types	**2.0** for all building types	**2.0** for all building types
8 / 400	9 / 200	10 / 200

Sub total for all building types in each category of sustainability

Divide by the number of building types for final total for buildings with a BREEAM rating

Final Total =

| 11 / 400 | 12 / 200 | 13 / 200 |

4.2 Building types not covered by BREEAM

Objective
- To make a commitment to achieve individual environmental targets for building types not covered by BREEAM.

Where the building type is not covered by BREEAM, the setting of targets for sustainability is more complicated. Rather than having a single rating or label that covers all the relevant issues it is necessary to set separate targets for each one. Some general suggestions on how to specify such targets are detailed below.

Questions

(a) What are the CO_2 targets for the proposed buildings?
Which thermal model will be used to demonstrate that the building has been designed to meet these targets?

Answering the Questions

(a) CO_2 emissions targets (energy targets)

This checklist uses targets expressed in terms of CO_2 emissions (Kg of carbon per annum), rather than energy targets (KWh per annum).

For non-domestic buildings one source of CO_2 emission targets is the UK Government's Energy Efficiency Best Practice programme (EEBPp) guides which cover a wide range of building types (see list below and *Useful contacts* box). Most of the guides contain energy and CO_2 figures representing a range of consumption from 'Typical' to 'Good Practice'.

This checklist recommends setting targets that are higher than the 'Good Practice' figures quoted in these guides to reflect the fact that the booklets cover a sample of buildings of different ages. The two levels proposed are 10% better than 'Good Practice' and 20% better than 'Good Practice'.

Demonstrating compliance with a target requires some prediction of the performance of the buildings under design. Modelling the performance of the building using computer software (known as thermal modelling) is the most common method of doing this. Several well-respected thermal models exist including SAP (Standard Assessment Procedure) for dwellings, TAS, Seri Res and ESP for commercial premises.

Rarely can the modelling be carried out at the early planning stage as too few of the details about the building exist at this stage. The checklist therefore asks the developer to commit to meeting the targets and to specify the details of the thermal model to be used.

EEBPp guides and Energy Consumption Guides

EEB 1   *Introduction to energy efficiency in schools*
EEB 3   *Introduction to energy efficiency in shops and stores*
EEB 4   *Introduction to energy efficiency in health care buildings*
EEB 5   *Introduction to energy efficiency in further and higher education buildings*
EEB 6   *Introduction to energy efficiency in offices*
EEB 7   *Introduction to energy efficiency in sports and recreation centres*
EEB 8   *Introduction to energy efficiency in libraries, museums and churches*
EEB 9   *Introduction to energy efficiency in hotels*
EEB 10 *Introduction to energy efficiency in post offices, building societies, banks and agencies*
EEB 11 *Introduction to energy efficiency in entertainment buildings*
EEB 12 *Introduction to energy efficiency in prisons, emergency buildings and courts*
EEB 13 *Introduction to energy efficiency in factories and warehouses*

Energy Consumption Guide 13   *Energy efficiency in public houses*
Energy Consumption Guide 15   *Saving energy in schools*
Energy Consumption Guide 18   *Energy efficiency in industrial buildings and sites*
Energy Consumption Guide 19   *Energy efficiency in offices*
Energy Consumption Guide 36   *Energy efficiency in hotels*

4 Impact of individual buildings

4.2 Building types not covered by BREEAM

Targets for the following issues are suggested:
- CO_2 emissions (energy targets),
- materials targets (use of green materials),
- water consumption targets,
- health and wellbeing (targets for daylighting, noise and fresh air provision),
- transport (provision of facilities for cyclists and runners),
- pollution (refrigerants and NO_X emissions targets).

CO_2 emissions targets

(a) For each building type enter the details of the thermal model to be used to predict CO_2 emissions in box 1. Calculate % of buildings of each type predicted to perform better than Good Practice CO_2 targets taken from DETR guides. Enter this in boxes 2 and 4. Multiply this by the scoring factor (SF) and enter the results in boxes 3 and 5. Enter the sum of boxes 3 and 5 in box 6. Multiply by the relevant sustainability factors to give the three separate scores in boxes 7, 8 and 9.

Building type 1

Totals

Building type 2

Totals

Building type 3

Totals

Building type 4

Totals

Sum of separate categories for each building type

Divide by the number of different building types for final total in each category

Range of performance

	Good practice	Best practice
Thermal model to be used, eg TAS, ESP, Seri Res.	10% better than DETR Good Practice: SF = 0.3	20% better than DETR Good Practice: SF = 0.5
	% meeting target × 0.3	% meeting target × 0.5

Suggested maximum scores for achieving Best Practice

Env.	Social	Econ.
3.0	1.0	1.0

Boxes numbered 1–5 repeat for each building type; totals in box 6; category totals in boxes 7, 8, 9.

Sum boxes: 10, 11, 12.

Final totals: 13 = 150, 14 = 50, 15 = 50.

Cont'd......

4.2 Building types not covered by BREEAM (cont'd)

Objective
- To make a commitment to achieve individual environmental targets for building types not covered by BREEAM.

(b) Use of green materials
Objective
- To minimise the damage caused to the environment by the use of low environmental impact materials and to ensure that timber is purchased from a sustainable source.

Questions

(b) What is the predicted use of low environmental impact building materials: (i) for construction elements?

Answering the Questions

(b) Green materials
(i) Construction elements chosen for low environmental impact
BRE has produced detailed guidance on the relative environmental merits of different construction types. This information is presented in two books called *The Green Guide to Specification* for commercial buildings and *The Green Guide to Housing Specification*. The Green Guides score the environmental impact of each type of construction presented on a scale of 'A' to 'C'. BREEAM credits those buildings where 80% (by area) of the main building elements score an 'A' rating. For building types not covered by BREEAM this standard is still appropriate. The main elements in BREEAM are defined as:
- upper floor slab,
- external walls,
- roof,
- windows.

For each type of building the % meeting the standard above (for all four elements) should be multiplied by the relevant factor to gain the scores in all three categories. For the final score, divide by the number of different building types.

4.2 Building types not covered by BREEAM (cont'd)

(i) Construction elements chosen for low environmental impact	Range of performance			Suggested maximum scores for achieving Best Practice		
	Minimum acceptable	Good practice	Best practice	Env.	Social	Econ.
(b) Calculate the % of buildings (for each building type) meeting the BREEAM criteria for green materials, ie with the four main construction elements meeting an 'A' rating in the relevant BRE *Green Guide*. Multiply by the relevant factor to gain the points score.	60–80% of buildings of this type	80–90 % of buildings of this type	> 90% of buildings of this type	0.4	0	0.25
Building type 1				/ 40	N/A	/ 25
Building type 2				/ 40	N/A	/ 25
Building type 3				/ 40	N/A	/ 25
Building type 4				/ 40	N/A	/ 25
Sub total for all building types in each category of sustainability				16	17	18
Divide scores in boxes 16, 17 and 18 by the number of different building types for final total in each category				19 / 40	20 / 0	21 / 25

Cont'd......

4.2 Building types not covered by BREEAM (cont'd)

Objective
- To make a commitment to achieve individual environmental targets for building types not covered by BREEAM.

(b) Use of green materials

Objective
- To minimise the damage caused to the environment by the use of low environmental impact materials and to ensure that timber is purchased from a sustainable source.

Questions

(b) What is the predicted use of low environmental impact building materials: (ii) sustainable timber?

Answering the Questions

(b) Green materials

(ii) Timber from a sustainably managed source
All timber for key elements including structural timber, cladding, carcassing and internal joinery is specified to come from sustainably managed sources. Currently, forests accredited to UKWAS (UK Woodland Assurance Scheme) or FSC (Forest Stewardship Council) qualify as sustainably managed. Imported timber products remain difficult to assess, but organisations like the Timber Trade Federation are working to address certification needs. To calculate points multiply the % of each building type meeting the criterion by the relevant factor. For the final score, divide by the number of different building types.

4 Impact of individual buildings

4.2 Building types not covered by BREEAM (cont'd)

(ii) Sustainable timber	Range of performance			Suggested maximum scores for achieving Best Practice		
	Minimum acceptable	Good practice	Best practice	Env.	Social	Econ.
(b) Calculate the % of buildings (for each building type) meeting the criteria for sustainable timber.	60–80% of buildings of this type	80–90 % of buildings of this type	> 90% of buildings of this type	0.4	0	0
Building type 1				/ 40	N/A	N/A
Building type 2				/ 40	N/A	N/A
Building type 3				/ 40	N/A	N/A
Building type 4				/ 40	N/A	N/A
Sub total for all building types in each category of sustainability				22	23	24
Divide by the number of different building types for final total in each category				25 / 40	26 / 0	27 / 0

Cont'd……

4.2 Building types not covered by BREEAM (cont'd)

Objective
- To make a commitment to achieve individual environmental targets for building types not covered by BREEAM.

(c) Water targets
Objective
- To save water resources.

Questions	Answering the Questions
(c) What is the predicted water consumption?	**(c) Water targets** Water consumption targets for those building types not covered by BREEAM are not yet available but the developer could state the water-saving measures as detailed opposite. Community-wide grey water recycling is covered in Section 5.3 *Water conservation*.

(d) Health and wellbeing
Objective
- To ensure that the internal environment of buildings is conducive to good health and productivity.

For a range of different building types it may be appropriate to ask for high standards of daylight provision and noise attenuation.

Questions	Answering the Questions
(d) *(i)* What % of buildings demonstrate a strong consideration of designing for maximum daylighting?	**(d) Health and wellbeing** *(i) Daylighting* The appropriate standards for these issues will vary considerably depending upon the type of building. At the moment (other than the standards built into BREEAM) there are no published set standards for these supported by the UK Government, other than the minimum requirement in building regulations. However, much research has been carried out on these topics and is referenced below. Advice on how to lay out the buildings on a site to gain maximum daylighting is available from BRE (see *References*).
(ii) What % of buildings have been designed to minimise noise attenuation between buildings and within buildings (to a standard higher than current building regulations)?	*(ii) Noise attenuation* At the time of printing this checklist, Part E of the Building Regulations was due to be published (see DTLR website for information).

4 Impact of individual buildings

4.2 Building types not covered by BREEAM (cont'd)

(c) Water targets	Range of performance			Suggested maximum scores for achieving Best Practice		
	Minimum acceptable	Good practice	Best practice	Env.	Social	Econ.
(c) What measures have been adopted to reduce water consumption?	Low flush WCs	Low flush WCs + spray taps + auto leak detector	Low flush WCs + spray taps + auto leak detector + water recycling	40	40	40
	All buildings not covered by BREEAM meet this	All buildings not covered by BREEAM meet this	All buildings not covered by BREEAM meet Good Practice and some have water recycling	28 / 40	29 / 40	30 / 40

(d) Health and wellbeing	Range of performance			Suggested maximum scores for achieving Best Practice		
	Minimum acceptable	Good practice	Best practice	Env.	Social	Econ.
(i) Daylighting Multiply % by the relevant category factor (E, S, Ec) to obtain the points scores in the separate categories.	60–80% of buildings	80–90 % of buildings	> 90% of buildings	0.2	0	0
				31 / 20	32 / 0	33 / 0
(ii) Noise attenuation Multiply % by the relevant category factor to obtain the points scores.	45–59% of buildings	60–90 % of buildings	> 90% of buildings	0.2	0.5	0.25
				34 / 20	35 / 50	36 / 25

Cont'd......

4.2 Building types not covered by BREEAM (cont'd)

Objective
- To make a commitment to achieve individual environmental targets for building types not covered by BREEAM.

(e) Transport
Objective
- To encourage walking, running and cycling as a transport mode as an alternative to car use.

Questions	Answering the Questions
(e) Have facilities to encourage walking, running and cycling, been included in most buildings?	**(e) Transport** Has the provision of cycle sheds and changing rooms for most buildings been explicitly mentioned in the proposal? If it covers >90% of all the buildings, Best Practice has been met. If it covers 60–90% of buildings, Good Practice has been met.

(f) Pollution
Objective
- To reduce air pollution from building materials and boiler plant.

Questions	Answering the Questions
(f) *(i)* What % of buildings use insulation materials with zero ozone depletion potential (ODP)?	**(f)** *(i) Insulation materials with zero ODP* Insulation in main structural elements (walls, roofs and floors).
(f) *(ii)* What % of buildings use refrigerants with zero ozone depletion potential (ODP)?	**(f)** *(ii) Refrigerants with zero ODP* Refrigerants in the air-conditioning system.
(f) *(iii)* What % of buildings use low NO_x emitting burners in boiler plant?	**(f)** *(iii) NO_x* The use of low NO_x emitting boiler plant has been specified.

4.2 Building types not covered by BREEAM (cont'd)

(e) Transport	Range of performance			Suggested maximum scores for achieving Best Practice		
	Minimum acceptable	Good practice	Best practice	Env.	Social	Econ.
Translate the % directly into points for each category.	45–59% of buildings	60–90 % of buildings	> 90% of buildings	37 / 100	38 / 100	39 / 100

(f) Pollution	Range of performance			Suggested maximum scores for achieving Best Practice		
	Minimum acceptable	Good practice	Best practice	Env.	Social	Econ.
(i) Zero ODP insulation materials. Multiply % by the relevant factor to obtain points score, eg 60% (of buildings) × 0.2 (Env. factor) = 12 points.	60–80% of buildings	80–90% of buildings	> 90% of buildings	0.2 (Factor)	0 (Factor)	0 (Factor)
				40 / 20	41 / 0	42 / 0
(ii) Zero ODP refrigerants. Multiply % by the relevant factor to obtain points score.	60–80% of buildings	80–90% of buildings	> 90% of buildings	0.2 (Factor)	0 (Factor)	0 (Factor)
				43 / 20	44 / 0	45 / 0
(iii) NO_x. Multiply % by the relevant category factor to obtain the points scores in the separate categories.	60–80% of buildings	80–90% of buildings	> 90% of buildings	0.1 (Factor)	0 (Factor)	0 (Factor)
				46 / 10	47 / 0	48 / 0

Cont'd......

To calculate the total score for each category of sustainability sum the boxes numbered on previous pages, as shown.

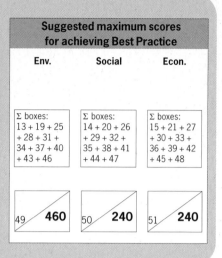

Suggested maximum scores for achieving Best Practice

Env.	Social	Econ.
Σ boxes: 13 + 19 + 25 + 28 + 31 + 34 + 37 + 40 + 43 + 46	Σ boxes: 14 + 20 + 26 + 29 + 32 + 35 + 38 + 41 + 44 + 47	Σ boxes: 15 + 21 + 27 + 30 + 33 + 36 + 39 + 42 + 45 + 48
49 / 460	50 / 240	51 / 240

Total for Environmental impacts of buildings without BREEAM ratings

Final total for Environmental impacts of buildings = points for buildings with BREEAM ratings (boxes 11, 12 and 13 from section 4.1) + points for buildings without BREEAM ratings (boxes 49, 50 and 51 from section 4.2)

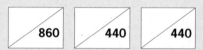

/ 860   / 440   / 440

Note: If all buildings are covered by BREEAM, double the final score for section 4.1. If none of the buildings are covered by BREEAM, double the score for section 4.2.

5 Natural resources

The natural resources used in construction in the UK are usually:
- *building materials,*
- *energy,*
- *water.*

The UK construction industry uses 6 tonnes of building materials per person each year. Of this, 20% is for infrastructure (civil engineering), while the remainder is for buildings. 250–300 million tonnes of material is extracted from quarries each year for aggregates, cement and bricks. Many other materials and components are used by the construction industry; each has a range of environmental consequences arising from its production, use, maintenance and final disposal.

In addition other natural resources can be damaged by construction resulting in:
- air pollution,
- water pollution,
- noise pollution,
- land pollution.

Land fill sites and quarries, for example, can cause all four types of pollution.

By reusing (reclaiming) and recycling building materials and choosing those with low environmental impacts, it is possible to minimise harmful environmental effects. By designing and building to minimise waste (known as 'lean construction') it is possible to reduce the amount of new building material wasted on site (currently 13 million tonnes of this goes straight to land fill sites).

Useful contacts

Environment Agency,
Rio House, Aztec West, Almondsbury, Bristol BS12 4UD

References

BRE. *Soakaway design.* Digest 365. Garston, CRC. 1991.
CIRIA. *Infiltration drainage: manual of good practice.* CIRIA Report 156. London, CIRIA. 1996.
Collins R J. *The use of recycled aggregates in concrete: BRE energy efficient office of the future.* Information Paper IP5/94. Garston, CRC. 1994.
DETR/DOE. Urban woodland and the benefits for local air quality. London, DETR/DOE. 1996.
Howard N, Edwards S & Anderson J. *BRE methodology for environmental profiles of construction materials, components and buildings.* BR 370. Garston, CRC. 1999.
Scottish Environment Protection Agency. *A guide to sustainable urban drainage.* Stirling, SEPA. 1997.
UK Database of Environmental Profiles of Construction Materials and Components. *www.bre.co.uk/envprofiles* (Internet subscription service).

5.1 Use of locally reclaimed/green materials

Objectives
- To reduce the amount of new building materials that we need to use on our developments and make best use of existing resources.
- To limit the amount of waste generated, either as land fill or by incineration.
- To ensure that locally available low environmental impact materials are specified and procured as widely as possible throughout the development.

The purpose of this section is to encourage:
- *the use of locally reclaimed material for roads, pavements and car parks,*
- *the specification of materials with low environmental impacts,*
- *the development of a green materials procurement policy.*

Note: The worthwhile distance involved in transporting reclaimed materials is dependent on the type of material; see ranking of materials in Table opposite.

Questions	Answering the Questions
(a) How much local reclaimed/green materials will be used for road construction?	(a) **Locally reclaimed materials for road construction** If a target of 10–20% (by volume or weight) has been set for the use of reclaimed material in roads, Good Practice has been met. If a target of more than 20% for recycled material has been set, Best Practice has been met.
(b) How much local reclaimed/green materials will be used for pavement construction?	(b) **Locally reclaimed materials for pavement construction** If a target of 10–20% (by volume or weight) has been set for the use of reclaimed material in pavements, Good Practice has been met. If a target of more than 20% for recycled material has been set, Best Practice has been met.
(c) How much local reclaimed/green materials will be used for car park construction?	(c) **Locally reclaimed materials for car park construction** If a target of 10–20% (by volume or weight) has been set for the use of reclaimed material in car parks, Good Practice has been met. If a target of more than 20% for recycled material has been set, Best Practice has been met.
(d) What effort has been made to specify locally available materials with low environmental impact?	(d) **Specification** If the proposal has been 'fully' considered, Best Practice has been met. If it has only been partially considered, Good Practice has been met.
(e) Does the developer have a policy of using local materials suppliers with environmentally friendly supply chains?	(e) **Procurement policy** If the procurement policy is comprehensive, Best Practice has been met. If it is only partial, Good Practice has been met.

5 Natural resources

5.1 Use of locally reclaimed/green materials

Maximum transport distances for reclaimed materials

Material	Distance (miles)
Reclaimed tile	100
Reclaimed slate	300
Reclaimed bricks	250
Recycled aggregates	150
Reclaimed timber (eg floor boards)	1000
Reclaimed steel products	2500
Reclaimed aluminium products	7500

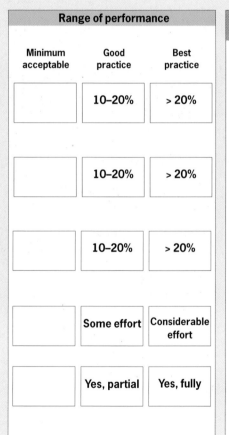

Range of performance			Suggested maximum scores for achieving Best Practice		
Minimum acceptable	Good practice	Best practice	Env.	Social	Econ.
	10–20%	> 20%	10	N/A	N/A
	10–20%	> 20%	10	N/A	N/A
	10–20%	> 20%	10	N/A	N/A
	Some effort	Considerable effort	50	25	25
	Yes, partial	Yes, fully	50	25	25

5.2 Air quality

Objective
- To minimise the air pollution from roads and buildings by planting trees and shrubs to help remove air-borne pollutants and particulates.

Questions

(a) What % of major roads/railway lines are screened from residential areas via belts of trees?

Answering the Questions

(a) Air quality
If more than 60% of the main roads are flanked by tree belts (or proposed tree belts), Best Practice has been achieved. If 30–60% of the main roads are flanked by tree belts, Good Practice has been achieved. Guidance on the design of woodland to achieve maximum air quality benefits is available in a DETR (DOE) report (see references).

5.3 Water conservation

Objectives
- To encourage water recycling in buildings and for landscaping purposes.

Despite the high level of rainfall in the UK, there are some parts of the UK (especially the South East) where water is scarce. The treatment of water for human consumption is expensive and resource intensive, and the demand for water has risen by 70% over the past 30 years. Water should therefore be conserved and recycled whenever practicable, and buildings should be individually metered to make the monitoring of water consumption straightforward.

Questions

(a) What % of recycled grey water is used in the development?

(b) What % of grey water/rainwater collection is used for landscaping purposes?

Answering the Questions

(a) Recycling grey water in buildings
Calculate the % of recycled grey water used in the development and multiply by 0.25.

(b) Recycling grey water for landscaping purposes
Calculate the % of grey water/rainwater collected which is used for landscaping purposes and multiply by 0.25.

5 Natural resources

5.2 Air quality

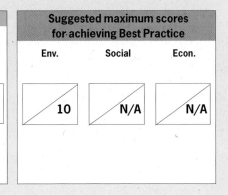

Range of performance			Suggested maximum scores for achieving Best Practice		
Minimum acceptable	Good practice	Best practice	Env.	Social	Econ.
10–30%	30–60%	60% +	10	N/A	N/A

5.3 Water conservation

Water-saving fittings should be adopted like spray taps, low flush WCs and leak detectors. Water-recycling systems that use grey water from sinks and baths to flush toilets, and water butts that collect water for landscaping are to be encouraged. When planning whole estates, it may be possible to plan in a grey water system that collects rainwater from a number of roofs and run-off points to be used for the general community.

Note: the water consumption of individual buildings via water-saving measures and water-efficient appliances is covered under BREEAM (for those buildings covered by BREEAM, see Chapter 4, section 4.1).

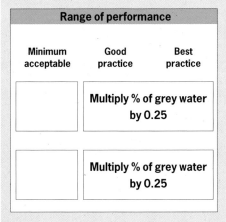
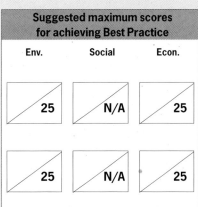

Range of performance			Suggested maximum scores for achieving Best Practice		
Minimum acceptable	Good practice	Best practice	Env.	Social	Econ.
	Multiply % of grey water by 0.25		25	N/A	25
	Multiply % of grey water by 0.25		25	N/A	25

64 A sustainability checklist for developments

5.4 Sustainable drainage

Objective
- To encourage surface water drainage systems that slow down the run-off rate to rivers/watercourses and aquifers, and so filter out some of the pollutants.

The design of surface water drainage should be considered at the earliest possible stages of the planning process so that the best management practices described in the Scottish Environmental Protection Agency's guide (see references) can be used. Ground conditions, in particular the permeability of the ground, needs to be considered as some geological formations are unsuitable for some of the sustainable options.

Questions

(a) Has a ground survey been done?

(b) What is the % of permeable paving in:
  (i) car parks?
  (ii) amenity areas?
  (iii) pedestrian pavements?
  (iv) other routes, eg cycle ways/bridleways?
  (v) swales and infiltration basins?

(c) What is the % of permeable conveyance systems (swales and filter drains)?

(d) What is the % of passive treatment systems (detention ponds, reed beds)?

Answering the Questions

(a) Ground survey
A ground survey is required to determine the suitability of the ground for sustainable drainage practices. If one has been carried out, Best Practice has been met.

(b) Surface control techniques
(i)–(iv) Assess the approximate proportion of permeable paving for all hard-standing areas and award the scores accordingly. If the percentage of permeable paving is high, the need for surface water drains is considerably reduced.

(v) Assess the proportion of the surface water run-off that is being captured in swales and infiltration basins. If this is > 50%, Good Practice has been met. If it is greater than 80%, Best Practice has been met.

(c) Permeable conveyance systems (swales and filter drains)
If conveyancing systems for surface water are being included, are they mostly permeable? If > 80% of the system is permeable, Best Practice has been achieved. If > 50% of the conveyancing systems are permeable, Good Practice has been met.

(d) Passive treatment systems
Where > 50% of the surface water is dealt with via passive treatment systems (like detention ponds and reed beds), Best Practice has been achieved. Where > 30% of the surface water is being dealt with via passive treatment systems, Good Practice has been met.

5.4 Sustainable drainage

	Range of performance		Suggested maximum scores for achieving Best Practice		
	Good practice	Best practice	Env.	Social	Econ.
		Yes	10	N/A	N/A
(i) Car parks	> 40%	> 50%	10	N/A	N/A
(ii) Amenity areas	> 40%	> 50%	10	N/A	N/A
(iii) Pavements	> 40%	> 50%	10	N/A	N/A
(iv) Other routes	> 40%	> 50%	10	N/A	N/A
(v) Swales and infiltration basins	> 50%	> 80%	10	N/A	N/A
	> 50%	> 80%	10	N/A	N/A
	> 30%	> 50%	10	N/A	N/A

5.5 Refuse composting

Objective
- To encourage recycling of green material by composting and chipping.

Local provision of composting facilities, or skips for taking green waste to central composting facilities, can reduce the waste going to land fill sites. In some cases, compost and wood chips can be given back to residents to re-use in their gardens, or can be used as part of the landscape management of public green space.

Questions

(a) How close are the nearest composting/chipping facilities for green material?

Answering the Questions

(a) Provision of local composting/chipping facilities
If a local composting or chipping centre is within 500 m of the centre of the development, Good Practice has been achieved. If there is provision for local composting on the proposed development site, within 200 m of the centre of the development, Best Practice has been met.

5.5 Refuse composting

Range of performance			Suggested maximum scores for achieving Best Practice		
Minimum acceptable	Good practice	Best practice	Env.	Social	Econ.
	< 500 m	< 200 m	10	N/A	10

Total for Natural resources 280 50 110

6 Ecology

The adverse impact of construction on wildlife and their habitats can be severe, sometimes wiping out whole species. This effect is usually a local one unless rare and endangered species are affected and then it can be considered to be global or national.

The ecological value of any site should be established by a survey carried out by a recognised expert. The Association of Wildlife Trust Consultancies (AWTC) can provide a list of qualified consultants. Measures to enhance the ecological value of a site can then be proposed in the development. These may include:
- landscaping,
- protecting the most important ecological attributes,
- creating new habitats and wildlife corridors, woodlands and wetlands.

Such features can provide pleasant amenity and leisure spaces for residents and enhance the quality of life.

Useful contacts

Association of Wildlife Trust Consultancies (AWTC),
Middlemarch Environmental Ltd,
Stoneleigh Deer Park, Staveton,
Kenilworth CV8 2LY

Council for the Protection of Rural England (CPRE),
Warwick House, 25 Buckingham Palace Road, London SW1W 0PP
Tel. 020 7976 6433;
Fax 020 7976 6373; www.cpre.org.uk

Joint Nature Conservation Committee,
Monkstone House, City Road,
Peterborough PE1 1JY
Tel. 01733 562626;
Fax 01733 555948; www.jncc.gov.uk

National Biodiversity Network,
The Kiln, Waterside, Mather Road,
Newark, Nottinghamshire NG24 1WT
Tel. 0870 0367711;
Fax 0870 0360101; www.nbn.org.uk

The Wildfowl and Wetlands Trust,
Barn Elms Lodge, Queen Elizabeth Walk,
Barnes, London SW13 0DB

The Wildlife Trusts,
The Kiln, Waterside, Mather Road,
Newark, Nottinghamshire NG24 1WT
Tel. 0870 0367711;
Fax 0870 0360101;
www.wildlifetrusts.org

References

Department of the Environment, Food and Rural Affairs. *Countryside survey 2000: assessing habitats in the UK countryside.* London, The Stationery Office. 2000.

Department of the Environment, Food and Rural Affairs. *Climate change and UK conservation: a review of the impact of climate change on UK species and habitat conservation policy.* London, The Stationery Office. 2000.

Department of the Environment, Transport and the Regions. *Nature conservation.* Planning Policy Guidance Note 9. London, The Stationery Office. 1994.

6.1 Conservation

Objective
- To maintain bio-diversity and protect natural habitats and features.

Questions

(a) Has a baseline survey of species, habitats and significant natural features been carried out?

(b) What percentage of all natural habitats have been preserved?

Answering the Questions

(a) Ecological survey
Has an ecological survey and evaluation of the site for proposed development been carried out by a recognised expert (eg someone recommended by the Association of Wildlife Trust Consultancies)? If the answer is Yes, partly, Good Practice has been met, if Yes, fully, then Best Practice has been met.

(b) Conservation
If the recommendations of the ecological survey regarding the preservation of natural habitats have been carried out for >70% of existing habitats, Best Practice has been achieved. If 60–70% of habitats have been preserved, then Good Practice has been met.

6.2 Enhancement of existing site

Objective
- To improve the ecological value of the site and provide new habitats and features.

Questions

(a) Has there been an increase in the natural habitats either by area or increased ecological value?

(b) Have any additional ecological features like woodland or wetland been included?

(c) Has a new wildlife corridor been added?

Answering the Questions

(a) Habitat creation/enhancement
If one type of habitat has been increased in area or quality, Good Practice has been met. If more than one type has increased in area or improved in quality, Best Practice has been achieved.

(b) Ecological features
Have any additional ecological features been added, like wetlands or woods? If one new ecological feature (excluding wildlife corridors that are dealt with below) has been added, Good Practice has been met. If more than one new ecological feature has been added, Best Practice has been met.

(c) Wildlife corridor
To qualify as a new corridor it must link in with an existing corridor or natural feature. Good Practice has been met if the corridor links in at one end, Best Practice has been achieved if it links in at both ends.

6 Ecology

6.1 Conservation

Range of performance			Suggested maximum scores for achieving Best Practice		
Minimum acceptable	Good practice	Best practice	Env.	Social	Econ.
	Yes, partly	Yes, fully	20	N/A	N/A
	> 60%	> 70%	20	10	N/A

6.2 Enhancement of existing site

Range of performance			Suggested maximum scores for achieving Best Practice		
Minimum acceptable	Good practice	Best practice	Env.	Social	Econ.
	Yes, in one habitat	Yes, in > 1 habitat	20	10	N/A
	One additional feature	> One additional feature	20	20	N/A
	Link at one end	Link at both ends	10	10	N/A

6.3 Planting

Objective
- To encourage the involvement of a qualified landscape architect to ensure that shrub and trees are planted to increase ecological value and provide variety and native species.

Questions

(a) Has expert advice (eg from a qualified landscape architect) been included in designing the development?

(b) Will the development significantly increase the number of trees in the area (after deducting any destroyed by development)?

(c) Has a mixture of native deciduous and evergreen trees and shrubs been specified?

Answering the Questions

(a) **Qualified landscape architect**
Has the advice of a qualified landscape architect been sought? If yes, Best Practice has been met.

(b) **Net increase in tree and shrub numbers**
Best Practice has been met if there is a more than three-fold increase in the number of trees and shrubs. Good Practice has been met if the number of trees and shrubs increases by a factor of 2 or more.

(c) **Native trees and shrubs**
If > 80% of the trees and shrubs specified are native species and represent a mixture of evergreen and deciduous varieties, Best Practice has been met. If 60–80% of the trees and shrubs specified are native species and represent a mixture of evergreen and deciduous varieties, Good Practice has been met.

6 Ecology

6.3 Planting

Range of performance			Suggested maximum scores for achieving Best Practice		
Minimum acceptable	Good practice	Best practice	Env.	Social	Econ.
		Yes	50	20	N/A
	2–3-fold increase	> 3-fold increase	50	N/A	N/A
	60–80% native	> 80% native	50	20	N/A

Total for Ecology — 240 | 90 | 0

7 Community

When the community in an area breaks down (for whatever reason) the area becomes undesirable and ultimately uninhabitable. Although lessons can be learned from the past in terms of planning for a sustainable society, there is no agreed formula for the built environment to ensure that it remains a 'good' place to live. Indeed, the management of an area after it has been built appears to have as much to do with keeping it desirable as the initial design.

In the White Paper *Our towns and cities*, the UK Government sets out its vision for the urban areas of the future. In section 3, the Paper outlines how this can only be achieved by allowing the local people to influence and inform the development process. Paragraph 3.1 states 'people have a right to determine their future and be involved in deciding how their town or city develops'.

The planning system obviously has a large part to play, but developers should get involved as early as possible, continuing their commitment throughout the construction/refurbishment phase and beyond if appropriate. The *Considerate Constructors Scheme* provides a creditable scheme and guidance on community affairs during construction.

In terms of design, there are well researched standards that have been proven to reduce crime and foster better social relations. The UK police have developed *Secured by Design,* a system that can be used to rate components, buildings and developments. A more general example is the *Urban design compendium* that gives advice on more thoughtful design.

Community management is a complex issue, with many actors having a role to play. The Social Exclusion Unit at the UK Government's Cabinet Office have issued *A new commitment to neighbourhood renewal* which highlights new policies and practices that aim to improve community functioning.

Useful contacts

Active Community Unit www.homeoffice.gov.uk/acu/acu.htm

Community Development Foundation www.cdf.org.uk

National Association of Councils for Voluntary Services www.nacvs.org.uk

National Council for Voluntary Organisations www.ncvo-vol.org.uk

Social Exclusion Unit www.cabinet-office.gov.uk/seu

References

Cabinet Office Social Exclusion Unit. *A new commitment to neighbourhood renewal: national strategy action plan.* London, The Stationery Office. 2001.
Considerate Constructors Scheme. www.ccscheme.org.uk.
DETR. *Our towns and cities: the future. Delivering an urban renaissance.* Government White Paper. London, The Stationery Office. 2000.
Llewellyn Davies. *Urban design compendium.* London, English Partnerships and The Housing Corporation. 2000.
Secured by Design. www.securedbydesign.com.

7.1 Community involvement and identity

Objectives
- To ensure that new development is supported by existing residents, integrates well with the existing community in terms of layout and design and, wherever practicable, enhances it.
- Where there is no existing community, a new development should have a distinct character and identity which will be present in quality proposals for both layout and design.

Questions	Answering the Questions
(a) Does the local community support the development?	**(a) Community support** Ways in which the new development can improve the quality of life for an existing community should be investigated. The community should be consulted about the proposed development and support obtained for the initial proposal. Community support could take the form of written support from local community and interest groups, eg parish councils, local councillors, local businesses and chambers of commerce, school governors, neighbourhood watch members, etc. If written support is obtained from many groups with little local objection, Best Practice has been met. If support significantly outweighs objections, Good Practice has been met.
(b) Is there a continuing programme of community involvement in the development plans?	**(b) Community involvement** Is there an ongoing programme of community involvement with the development, eg participation and feedback days, information leaflets, web-sites, events, etc. If a full and detailed programme is included then Best Practice has been met.
(c)(i) For a site with an existing community, will the development be well integrated into the existing identifiable community?	**(c) Integration with existing community or identity of new community** (i) Does the development integrate well with any existing development to ensure that it forms part of the existing community? If it integrates well in most respects (eg uses, access, boundaries, scale, relationship to local centre, etc.), Best Practice has been achieved. If the development only integrates well in some respects, Good Practice has been met.
(ii) For a site with no existing community, will the development create a new community with a strong identity?	(ii) If there is no existing community, does the new development have the potential to have a strong community identity due to the layout of the settlement, the provision of an identifiable centre and facilities, etc.? If this has been made a priority in the design, Best Practice has been achieved. If it has been considered in most respects but not all, Good Practice has been achieved.
(d) Does the development significantly enhance the existing area?	**(d) Enhancement of existing development** Does the new development enhance the existing area by adding more facilities, amenity space, areas of natural beauty, etc.? If the development enhances in more than one respect, Best Practice has been achieved. If the development enhances in one respect only, Good Practice has been achieved.
(e) Has a householder's pack with information on the following local services and community issues been provided? ● Public transport services ● Local facilities/amenities ● Energy efficiency ● Crime prevention ● Water conservation ● Refuse collection and composting	**(e) Information pack to homes and businesses** If a pack has been provided that covers at least 3 of the topics left, Good Practice has been achieved. If a pack has been provided that covers at least 5 of the topics left, Best Practice has been achieved.

7 Community

7.1 Community involvement and identity

Range of performance			Suggested maximum scores for achieving Best Practice		
Minimum acceptable	Good practice	Best practice	Env.	Social	Econ.
	Support > Resistance	Yes, strongly	N/A	50	50
	Partial	Extensive	N/A	50	50
	Partially	Yes, well	N/A	20	N/A
	Partially	Significantly	N/A	20	N/A
	Partially	Significantly	N/A	20	20
	Yes, with 3 topics listed	Yes, with 5 or more topics listed	20	20	20

7.2 Measures taken to reduce the opportunity for crime

Objective
- To design developments to reduce the opportunity for crime and provide a safe environment for residents wherever possible.

Questions	Answering the Questions
(a) What % of housing has been designed to 'Secured by design' standards?	**(a) Housing design** If > 80% of the housing is to be built to 'Secured by design' standards, Best Practice has been achieved. If 60–80% of the housing is designed to meet this standard, Good Practice has been achieved.
(b) Has advice been sought from the police or other recognised expert body on estate layout?	**(b) Estate layout** If police advice has been sought (or will it be sought) on the layout of the estate, Best Practice has been achieved.
(c) What % of parking spaces and walkways have been designed to be 'overlooked' by housing or offices wherever possible?	**(c) Parking space and walkway design** If 30–60% of parking spaces and walkways have been designed to be overlooked, Good Practice has been met. If more than 60% have been designed to be overlooked, Best Practice has been achieved.
(d) What % of bus shelters are within 20 m of public telephones?	**(d) Public telephones** If over 80% of the bus shelters have public telephones within 20 m, Best Practice has been met. Good Practice has been met if 40–80% of the bus shelters have a public telephone within 20 m.
(e) What % of public places have security lighting and cameras?	**(e) Security lighting and cameras** If 30–50% of public places have security lighting and cameras, Good Practice has been achieved. If more than 50% of public places have security lighting and cameras, Best Practice has been met.

7.2 Measures taken to reduce the opportunity for crime

Range of performance			Suggested maximum scores for achieving Best Practice		
Minimum acceptable	Good practice	Best practice	Env.	Social	Econ.
	60–80%	< 80%	N/A	20	20
N/A	N/A	Yes	N/A	20	20
	30–60%	> 60%	N/A	20	20
	40–80%	> 80%	20	20	20
	30–50%	> 50%	N/A	10	10

Total for Community 40 270 230

8 Business

The economic prosperity of every region of the UK is the responsibility of the relevant Regional Development Agency. One of their statutory purposes is to further economic development and regeneration, as well as to contribute to sustainable development. To that end they are required to produce a regional strategy detailing the economic prospects for the region and the type of investment it needs to attract. Such strategies may also specify the area(s) earmarked for development in order to meet specific investment needs. In the regional strategy the Agency must also consider the four themes of sustainability covered in the UK Strategy for Sustainable Development:
- maintenance of high and stable levels of economic growth and employment,
- social progress which recognises the needs of everyone,
- effective protection of the environment,
- prudent use of natural resources.

During the development of their Strategies, Regional Development Agencies must consider the Regional Planning Guidance, Regional Transport Strategies, Supplementary Planning Guidance and Local Plans. Economic strategies should complement the work of the local Government Offices (GO) and local authorities in the area.

RDAs, GOs and local authorities also have a vital role to play in terms of training opportunities. Using resources provided by the *New deal for communities*, the Department for Education and Skills (DfES) Neighbourhood Support Fund and Single Regeneration Budgets, and the powers and duties associated with, for example, life-long learning and promoting community well-being, many opportunities for providing local training and employment can be exploited.

It is essential that activities arising from this area are well integrated with the social and environmental issues. Integration ensures 'win-win' situations, for example, when the solution for an environmental or social concern has economic benefits.

Outside England, economic development is covered by similar organisations in Wales, Scotland and Northern Ireland: the Welsh Development Agency, Scottish Enterprise and the Department for Regional Development, respectively.

Useful contacts

One North East www.onenortheast.co.uk
Northwest Development Agency www.nwda.co.uk
Yorkshire Forward www.yorkshire-forward.co.uk
Advantage West Midlands www.advantage-westmidlands.co.uk
East Midlands Development Agency www.emda.org.uk
East of England Development Agency www.eeda.org.uk
South West of England Development Agency www.southwestrda.org.uk
South East of England Development Agency www.seeda.co.uk
London Development Agency www.lda.gov.uk
Scottish Enterprise www.scottish-enterprise.com
Welsh Development Agency www.wda.co.uk
Northern Ireland Department for Regional Development www.drdni.gov.uk

8.1 Enhanced business opportunities

Objective
- To demonstrate that the proposed development (whether new build or regeneration) will bring economic prosperity to the local community.

Partnerships between the RDA, local authority, local business organisations such as the Chamber of Commerce, and voluntary organisations can be particularly beneficial at this level.

Questions | Answering the Questions

(a) Does the proposed development meet the general requirements of the economic strategy?

(a) Meeting the up-to-date economic strategy
If the proposal meets the guidance laid down in the economic strategy fully, Best Practice has been achieved. If it largely meets the strategy, Good Practice has been met.

(b) How great is the ability of the development to attract inward investment?

(b) Inward investment
If the development has strong potential to attract inward investment (eg there are already some businesses that wish to relocate behind the proposal), Best Practice has been achieved. If the proposal has potential to attract significant inward investment (but none can be quantified or named at the planning proposal stage), Good Practice has been achieved.

(c) Will the development increase the business base in the area?

(c) Business base
Is the new development likely to diversify the business base of the area? If it has strong potential to do so, Best Practice has been met. If the development has reasonable potential to do so, Good Practice has been met.

(d) Will the development help to maintain property values in and close to the development?

(d) Property values
Is the development likely to maintain the property values in and around the development? If it will affect only those in the development, Good Practice has been met. If it maintains the property values around the development as well, Best Practice has been achieved.

(e) Will the development result in increased viability of existing businesses and public transport?

(e) Viability of existing businesses
If the development is likely to increase the viability of both existing businesses and public transport, Best Practice has been achieved. If it is likely to improve the viability of one or the other, Good Practice has been achieved.

8.1 Enhanced business opportunities

Range of performance			Suggested maximum scores for achieving Best Practice		
Minimum acceptable	Good practice	Best practice	Env.	Social	Econ.
	Yes, partly	Yes, well	N/A	10	10
	Medium ability	Great ability	N/A	50	100
	Yes, partly	Yes, well	N/A	50	100
	In development only	In and around development	N/A	25	50
	One only	Both	20	20	50

8.2 Employment and training

Objective
- To enhance the job and training prospects for local people.

Training opportunities can prove an effective means of entering the job market for people who are long-term unemployed or who do not have many academic qualifications. Training in specific areas like construction, the environment, or community services may be particularly beneficial, addressing skill shortages that could delay the development, and ensuring that local people are trained to fill vital community management posts.

Questions

(a) What is the ability of the development to create permanent jobs?

(b) Are there any proposals to train unemployed local people as part of the development process?

(c) Will any new jobs be created which will protect/manage the environment?

(d) Will any new jobs be created which will maintain or enhance the social aspects of the community?

Answering the Questions

(a) Permanent jobs creation
Assess the ability of the development to create permanent jobs (largely for local people). If the ability to create jobs is high, Best Practice has been met. If the ability is medium, Good Practice has been met.

(b) Training opportunities
If there are significant training opportunities for local unemployed people, Best Practice has been met. If there are valuable but not extensive opportunities, Good Practice has been met.

(c) Jobs protecting the environment
The number of jobs created by a development will depend upon its size and nature. This aspect should therefore be tailored to suit the circumstances. The recommendations below are rough guidelines only.
If 1–2 jobs are likely to be created which will protect/manage the environment, Good Practice has been met. If more than 3 jobs are likely to be created to protect/manage the environment, Best Practice has been met.

(d) Jobs enhancing the community
The number of jobs created by a development will depend upon its size and nature. This aspect should therefore be tailored to suit the circumstances. The recommendations below are rough guidelines only.
If 1–2 jobs are likely to be created which will enhance the community, Good Practice has been met. If more than 3 jobs are likely to be created to enhance the community, Best Practice has been met.

8.2 Employment and training

Range of performance			Suggested maximum scores for achieving Best Practice		
Minimum acceptable	Good practice	Best practice	Env.	Social	Econ.
	Medium	High	N/A	100	100
	Some	Extensive	N/A	50	20
	1–2 jobs	3+ jobs	20	20	20
	1–2 jobs	3+ jobs	20	40	20

Total for Business issues — 60 / 365 / 470

Calculating the overall scores for social, environmental and economic sustainability

If a summary of the scores for a development is considered helpful this can be calculated and presented in a number of ways, three examples of which are given below.

1 Scores for each of the eight main issue headings
Scores for environment, social and economic categories can be calculated for each issue heading and can be presented as a fingerprint diagram (see below). In addition, the actual scores awarded can be presented alongside the maximum scores.

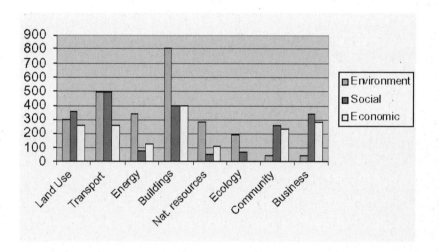

Scoring for each issue heading

2 Total scores for comparing proposals
All the scores in the three categories can be totalled to gain three overall scores. However, when comparing one proposal with another, or with the requirements of a plan, the separate theme fingerprint (see next example) provides much more detail on strengths and weaknesses.

3 Scores for sub issues within issue headings
If further detail is required the fingerprint could show each sub issue (see next page).

Calculating the overall scores

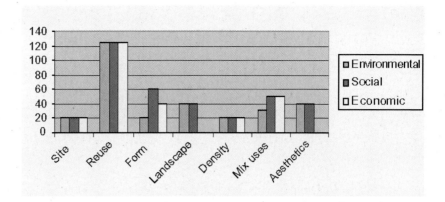

Scoring for each sub issue within an issue heading, eg Land use, urban form and design

Weighting and ranking of the issues

Other than the weightings implicit in the recommended maximum scores, no further weightings are suggested in this checklist.

Weighting different aspects of sustainability is always contentious and until 1998 BRE did not attempt to do this. However, research carried out by the *Centre for Sustainable Construction* at BRE in 1997/1998 established a preliminary set of consensus weightings for the environmental, social and economic aspects of sustainability based on the priorities and views of a wide range of interest groups. This work was funded under the Department of the Environment, Transport and the Region's *Partners in Technology* programme. The project found a surprisingly high degree of consensus about the relative importance of different sustainability issues across the following interest groups:
- Government policy makers,
- Construction professionals,
- Local authorities,
- Materials producers,
- Developers and investors,
- Environmental groups and lobbyists,
- Academics.

The results of this project are presented in BRE *Digest 446* (see details below). The suggested scores in this checklist have been influenced by the results of the Ecopoints exercise. However, this study will be updated regularly by BRE to reflect changing views and priorities. Similarly, the user of the checklist must ensure that the priorities implicit in the scoring are appropriate for the local conditions. For example, in areas of economic boom, environmental and social factors could be weighted as more important than economic issues. In areas of economic decline, social and economic factors are crucial to the success of any development and so may be valued more highly than environmental issues.

Reference
Dickie I & Howard N. *Assessing environmental impacts of construction: Industry consensus, BREEAM and UK Ecopoints.* BRE Digest 446. Garston, CRC. 2000.